Mathematics Behind Fuzzy Logic

Esko Turunen

Mathematics Behind Fuzzy Logic

Physica-Verlag
A Springer-Verlag Company

Dr. Esko Turunen
Senior Researcher
Department of Information Technology
Lappeenranta University of Technology
P.O. Box 20
FIN-53851 Lappeenranta
Finland
E-mail: esko.turunen@lut.fi

ISBN 3-7908-1221-8 Physica-Verlag Heidelberg New York

Cataloging-in-Publication Data applied for
Die Deutsche Bibliothek – CIP-Einheitsaufnahme
Turunen, Esko: Mathematics behind fuzzy logic/Esko Turunen. – Heidelberg: Physica-Verl., 1999
(Advances in soft computing)
ISBN 3-7908-1221-8

Printed in Germany

Softcover Design: Erich Kirchner, Heidelberg

SPIN 10730542 88/2202-5 4 3 2 1 0 – Printed on acid-free paper

Preface

Logic is art of inference. Having a set of true statements, we are interested in their logical consequences which we regard to be true, too. Mathematicians have applied two valued logic thousands of years with great success for the problems rising in the rigorous world of mathematics. For example, from true premises *4 is an even number* and *If n is an even number, then n+2 is an even number* they conclude correctly that *6 is an even number*, therefore *8 is an even number*, and so on. The real world in which we are living is not, however, rigorous like the artificial world of mathematics, but is full of vagueness and uncertainty. Two valued logic fits well to mathematical reasoning, in some sense two valued logic is the metamathematics of mathematics. Applying two valued logic outside mathematics, however, rise anomalies that we cannot accept as they contradict our everyday experiences. As an example, consider all the languages and dialects spoken in the modern world. Definitely, there are languages that are very close to each other; Czech and Slovak, for instance are such languages while English and Chinese are not. Similarly, the answer to a question *Does a person who understands the Swedish dialect spoken in Finland also understand the Swedish spoken in the capital of Sweden?* is affirmative if one has only two choices 'yes' or 'no'. In a same manner, a person understanding the Swedish spoken in Stockholm also understands the Swedish dialect spoken in the southern Sweden, and a person who understands the southern Sweden dialect understands Danish, too. Thus, a conclusion by two valued logic is that a person understanding the Swedish dialect spoken in Finland would also understand Danish, which is, however, quite questionable.

One alternative approach to avoid this and other similar paradox is to accept more truth values that *true* and *false* only. This is the starting point of many-valued logics and fuzzy logic. Following Lotfi Zadeh, the inventor of fuzzy sets and fuzzy logic, we distinguish fuzzy logic in *broad sense* i.e., everything concerning vagueness and fuzziness, from fuzzy logic in *narrow sense*, that is, the formal logical calculus of fuzziness. In this book we focus on the latter.

Jan Łukasiewicz, a Polish mathematician, was the first to investigate systematically many-valued logics in 1920's. In 1935, Morchaj Wajsberg showed that infinite valued propositional logic was complete with respect to the axioms conjectured by Łukasiewicz. Twenty-three years later, in 1958, C.C. Chang introduced MV-algebras, which allowed him to give another completeness proof for Łukasiewicz logic. For decades many-valued logic was far from the mainstreams of mathematical research, it was only after the 'fuzzy boom' started in 1965 with Zadeh's seminal paper 'Fuzzy Sets' that the situation has changed a bit. In 1979 Czech Jan Pavelka published a paper entitled 'On fuzzy logic' in which he generalized Łukasiewicz's logic by introducing fuzzy consequence operations, general fuzzy rules of inference, fuzzy proofs, etc. Pavelka studied the real unit interval valued fuzzy propositional logic and proved that necessary and sufficient condition for the completeness of his logic is the continuity of the implication operation. In 1989 another Czech Vilem Novák extended Pavelka's results on [0,1]-valued fuzzy predicate logic. The contribution of the present author was to show in 1994 the algebraic connections of Pavelkas's logic, e.g. that in the real unit interval Pavelka style fuzzy logic is axiomatizable if, and only if the set of truth values forms a complete MV-algebra of [0,1].

Many results in fuzzy logic depend on the mathematical structure the truth value set obeys. In this textbook we study systematically the algebraic foundations of many-valued and fuzzy reasoning in general, and Pavelka's fuzzy logic in particular.

The book is self-containing, thus no previous knowledge in algebra or in logic is required. The reader should, however, be familiar with standard mathematical reasoning and denotation. The book contains many exercises varying from routine computations to more sophisticated theoretical ones. All the exercises, however, are within the scope of the text. Although some details of the text are left for the reader as exercises, the main results of the book are independent of these exercises. As the exercises play an important role in learning the material, we highly recommend the reader to do them; complete answers can be found at the end of the book.

In Chapter 1 we study such basic notions as order, lattice, residuum and bi-residuum, which turn out to be fundamental in many applications. Residuated lattices were studied already in 1930's by Dilworth and Ward, and we see how they are related to the famous French mathematical Everiste Galois's ideas. Section 1.4 contains results on BL-algebras, a structure invented by Petr Hájek, Czech Republic. This material offers an introduction to Hájek's book 'Metamathematics of Fuzzy Logic', which we recommend for further reading.

Chapter 2 contains all the necessary algebraic results we need to be able to prove in details a general completeness theorem of Pavelka's logic. We study Wajsberg algebras and MV-algebras, and show their mutual equivalence. MV-algebras are particular residuated lattices, however, from application point of view they possess the best properties as we will see. The results we study are due to J.M. Font, A.J. Rodriques, A. Torres, E. Trillas and L. Valverde, Spain, Peter Belluce, Canada, late Daniel Gluschankof, Argentina, Antonio di Nola, Salvatore Sessa, L. Lacava, P. Mangani and, of course, Daniele Mundici, Italy. For further reading on MV-algebras we recommend 'Algebraic Foundations of Many-valued Reasoning' a book by R. Cignoli, I.M.L. D'Ottaviano and D. Mundici to appear in the near future.

In Chapter 3 we focus in details on Pavelka's fuzzy sentential logic in the algebraic form it was studied by the present author. Besides the deep theoretical results such as completeness, we give several examples on how to apply Pavelka's logic.

Chapter 4 concerns fuzzy relations. First we give necessary and sufficient conditions for the existence of solutions of general fuzzy relation equations, proved by the author, and then we study fuzzy similarity and its connection to approximate reasoning, due to Frank Klawonn, Germany, and Juan Luis Castro, Spain. It is well-known that, from mathematical point of view, various defuzzication methods of approximate reasoning are problematic as they seem to have no logical foundations. In section 4.4. we overcome this problem by introducing a new method for many-valued reasoning based on fuzzy similarity. It is of importance to realize that our method does not require any defuzzification technique. We conclude the textbook by presenting some real world applications which exploit our method. As usual in mathematical text, we abbreviate iff for if, and only if.

Acknowledgement. The author is very indebted to Vilem Novák, Petr Hájek, Daniele Mundici, Antonio di Nola, Ulrich Höhle and other participants of COST 15 Action[1] for fruitful discussions on the subject of the present textbook. Special thanks are due to Salvatore Sessa for careful reading the manuscript and many helpful comments on how to improve it. Last, but not least, thanks to my diploma thesis student Mika Hempilä, who inspired me to start to study applications of fuzzy similarity.

Esko Turunen
9.5.1999, Riihimäki, Finland

[1] Many-valued Logic for Computer Science Applications

Contents

Chapter 1

Residuated Lattices

1.1 Lattices and Equivalence Relations

We start our algebraic study on fuzzy logic by introducing such concepts as order, equivalence relation, lattice, etc. These concepts will play an important role when later giving a relevant mathematical structure for the set of values of truth and proving deeper properties of fuzzy logic. Order, for example, in an essential concept in fuzzy logic as our fundamental assumption will be that truth can have a degree varying from total false to absolute true. Moreover, we want that various truth values are in at least come level comparable. Our assumption is that the reader is familiar with such fundamental mathematical objects as set, void set, function, product set, etc. which we do not define here.

Let A be a non-void set. A *binary relation* R on A is a subset of the product set $A \times A$; if $(x, y) \in R$ we denote xRy. A binary relation R is a *quasi-order* on A if it is *reflexive* and *transitive*, i.e. for all $x, y, \in A$,

$$xRx, \tag{1.1}$$

$$\text{if } xRy \text{ and } yRz, \text{ then } xRz. \tag{1.2}$$

A quasi-order R on A is an *(partial) order* on A if it is also *anti-symmetric*, i.e. for all $x, y, \in A$,

$$\text{if } xRy \text{ and } yRx, \text{ then } x = y. \tag{1.3}$$

A quasi-order R on A is an *equivalence relation* on A if it is *symmetric*, i.e. for all $x, y, \in A$,

$$\text{if } xRy, \text{ then } yRx. \tag{1.4}$$

Equivalence relation is usually denoted by $\sim$ and quasi-order which is not symmetric is denoted by $\leq$.

Example 1 Let A be the set of all human beings. Define a binary relation R on A by xRy if x is at least as tall as y. Clearly R is reflexive and transitive but it is not anti-symmetric; for sure there exist two persons with equal height. Obviously, R is not symmetric, either.

Example 2 Let $A = \{x\}$ and R the only non-void binary relation on A, that is xRx. Then R is both symmetric and anti-symmetric.

Example 3 Assuming every human being is of exactly one nationality, define a binary relation R on the set of all human beings by xRy if x and y are of the same nationality. Then R is an equivalence relation and is not an order relation.

Example 4 The natural order on the set of real numbers is an example of a binary relation which is an order and is not an equivalence.

If $\leq$ is an order on A, $x \leq y$ and $x \neq y$ we write $x < y$.

Example 5 A set of truth values

$$\{ \text{ false, more-or-less-false, unknown, more-or-less-true, true } \}$$

can be ordered in a natural way by setting

$$\text{false} < \text{more-or-less-false} < \text{unknown} < \text{more-or-less-true} < \text{true}.$$

Let $\sim$ be an equivalence relation on A. The sets

$$|x| = \{y \in A, y \sim x\} \text{ , where } x \in A,$$

are called the *equivalence classes* of the equivalence relation $\sim$. The set of all equivalence classes $|x|$, $x \in A$, is denoted by $A/\sim$.

Remark 1 *Let $\sim$ be an equivalence relation on A. Then for every $x, y \in A$,*
(a) $x \in |x|$,
(b) $x \sim y$ *iff* $|x| = |y|$ *iff* $x \in |y|$,
(c) *if $x \sim y$ does not hold, then the classes $|x|$ and $|y|$ are disjoint.*

Proof. Exercise.

Assume $\sim$ is an equivalence relation on A. Let f be an m-ary operation on A, i.e. a mapping $f : A^m \mapsto A$, m a natural number. If conditions

$$a_1 \sim b_1, \cdots, a_m \sim b_m \tag{1.5}$$

imply

$$f(a_1, \cdots, a_m) \sim f(b_1, \cdots, b_m) \tag{1.6}$$

we say that $\sim$ is a *congruence* with respect to f. If $\sim$ is a con with respect to an operation $f : A^m \mapsto A$, then by writing

$$g(|a_1|, \cdots, |a_m|) = |f(a_1, \cdots, a_m)| \tag{1.7}$$

we define an m-ary operation on $A/\sim$. This definition is correct since, by (1.6), the right hand side of (1.7) does not depend on the choice of representatives $a_1, \cdots, a_m$ of the equivalent classes $|a_1|, \cdots, |a_m|$.

Proposition 1 *Let R be a quasi-order on a non-void set A. For each $x, y \in A$, write*

$$xEy \text{ iff } xRy \text{ and } yRx. \tag{1.8}$$

Then the binary relation E is an equivalence relation on A. Moreover, for each $x, y \in A$, let

$$|x|S|y| \text{ iff } xRy. \tag{1.9}$$

Then the binary relation S is an order on A/E.

Proof. The reflexivity of E follows from (1.1). Let xEy, yEz. Then xRy, yRz and zRy, yRz, therefore xRz, zRx. Thus xEz, i.e. E is transitive. The symmetry of E follows from (1.8). Thus E is an equivalence relation on A. We have to show that definition (1.9) does not depend on the choice of representatives x, y of the sets $|x|$ and $|y|$, respectively. Indeed, if xRy and $z \in |x|$, $w \in |y|$, then xEz, yEw and, by (1.8), zRx, yRw. Now zRx, xRy, yRw implies zRw, i.e. (1.9) is correct. To show that (1.9) defines an order on A/E we first realize that S is reflexive and transitive as R is. Moreover, if $|x|S|y|$ and $|y|S|x|$, then xRy and yRx. Hence xEy, and therefore $|x| = |y|$. This shows that S is anti-symmetric on A/E. □

A ***partially ordered set*** or ***poset*** is a set A on which an order relation $\leq$ has been defined. Of course, on a set A various order relations can be defined. If in a poset A either $x \leq y$ or $y \leq x$ for each $x, y \in A$, then A is *linear* and is called *a chain.* In such case the order $\leq$ is *total order* and A is called *linearly ordered.*

Example 6 The set $\mathcal{R}$ of real numbers endowed by the usual order relation is a chain as well as the *unit interval* $I = \{0 \leq x \leq 1 \mid x \in \mathcal{R}\}$ under the same order. By defining on the product set $I \times I$, for each $(a, b), (c, d) \in I \times I$,

$$(a, b) \leq (c, d) \text{ iff } a \leq c \text{ and } b \leq d$$

we obtain an example of a poset which is not a chain, e.g. $(0, 1)$ and $(1, 0)$ are *incomparable*; $(0, 1) \not\leq (1, 0)$ and $(1, 0) \not\leq (0, 1)$.

Example 7 The family E of *open sets* of the real line $\mathcal{R}$ is defined in the following way: (a) $\emptyset \in E$ (b) any open interval is in E (c) any union of open sets in E is in E (d) the intersection of two open sets in E is in E. The set theoretical inclusion $\subseteq$ is an order on E, however, E is not a chain under this order.

An element a of a poset A is said to be *maximal* if there is no element $b \in A$ such that $a \leq b$ and $a \neq b$. Dually, an element a is said to be *minimal* if there is no such element $b \in A$ such that $b \leq a$ and $b \neq a$. A poset can contain many maximal and minimal elements. For instance, if the partial ordering is defined via $x \leq y$ iff $x = y$, then every element is simultaneously maximal and minimal.

An *upper bound* of a subset X of a poset A is an element $a \in A$ such that $x \leq a$ for every $x \in X$. The *least upper bound* (l.u.b) is an upper bound a such that, if $b \in A$ is an upper bound of X, then $a \leq b$. The l.u.b. of a set $X \subseteq A$ is denoted by $\bigvee\{x \mid x \in X\}$, in particular $x \vee y$ for $\bigvee\{x, y\}$. Similarly, a *lower bound* of $X \subseteq A$, if it exists, is an element $a \in A$ such that $a \leq x$ for every $x \in X$. A lower bound a is the *greatest lower bound* (g.l.b) if for every lower bound b of X $b \leq a$ holds.

The g.l.b. of a set $X \subseteq A$ is denoted by $\bigwedge\{x \mid x \in X\}$, in particular $x \wedge y$ for $\bigwedge\{x, y\}$.

The set $\mathcal{N}$ of natural numbers, for example, when considered as a subset of $\mathcal{R}$ has the greatest lower bound, namely 1, but does not have any upper bound.

We introduce an important

Zorn's Lemma If every chain of elements of a partially ordered set A has an upper bound in A, then A contains a maximal element. More precisely, for every $a \in A$, there exists a maximal element b such that $a \leq b$.

To prove this Lemma in details, we should study transfinite increasing sequences which, however, are all too far from the main scope of our present interest.

If in a poset A there exists an element a such that $a \leq x$ for all $x \in A$, then a is called the *least element* or *zero* of A and is denoted by $\mathbf{0}$. Dually, an element $b \in A$ such that $x \leq b$ for all $x \in A$, if it exists, is the *greatest element* or *unit* of A and is denoted by $\mathbf{1}$. Clearly, a poset A contains at most one least element and at most one greatest element. The set of natural numbers contains the least element but does not contain the greatest element, the set of real numbers does not contain the least element or the greatest element. The least element and the greatest element of the poset E of the open sets of the real interval are $\emptyset$ and $\mathcal{R}$, respectively.

Definition 1 *A* lattice *is a poset L such that for any $x, y \in L$, $x \wedge y$ and $x \vee y$ exist in L. $x \wedge y$ is called the* meet *of x and y and $x \vee y$ is called the* join *of x and y. A lattice L is a* (countable) complete lattice *if* $\bigvee\{x \mid x \in X\}$ *and* $\bigwedge\{x \mid x \in X\}$ *exist in L for any (countable) subset $X \subseteq L$.*

By setting $X = L$ we see that any complete lattice contains the least element $\mathbf{0}$ and the greatest element $\mathbf{1}$. The unit interval I, for example, is a complete lattice under the usual order and $x \vee y = \max\{x, y\}$, $x \wedge y = \min\{x, y\}$. The unit interval of rational numbers $I_Q = \{x \mid 0 \leq x \leq 1, x \in Q\}$ is a lattice under the same order as I. It is not, however, a complete lattice; for example

$$X = \{x \mid 0 \leq x < 1/\sqrt{2}, x \in Q\}$$

is a subset of I_Q but $\bigvee\{x \mid x \in X\} = 1/\sqrt{2} \notin I_Q$.

Define on the poset E of the open sets of the real line operations $\wedge = \cap$ (intersection) and $\vee = \cup$ (union). Then E becomes a lattice.

Remark 2 *Define on a chain A, for all $x, y \in A$, $x \wedge y = x$, $x \vee y = y$ iff $x \leq y$. Then we obtain a lattice.*

Proof. Exercise.

Remark 3 *Assume A is a poset. Then*

$$x \wedge x = x, x \vee x = x, \quad \text{idempotency} \quad (1.10)$$

$$x \wedge y = y \wedge x, x \vee y = y \vee x, \quad \text{commutativity} \quad (1.11)$$

$$x \wedge (y \wedge z) = (x \wedge y) \wedge z, x \vee (y \vee z) = (x \vee y) \vee z, \quad \text{associativity} \quad (1.12)$$

$$x \wedge (x \vee y) = x \vee (x \wedge y) = x, \quad \text{absorption} \quad (1.13)$$

$$x \leq y \text{ iff } x \wedge y = x \text{ iff } x \vee y = y, \quad \text{consistency} \quad (1.14)$$

whenever these equations exist in A.

Proof. Exercise.

Proposition 2 *In a lattice L, the operations meet and join are* isotone, *i.e. for any $x, y, z \in L$, if $y \leq z$ then $x \wedge y \leq x \wedge z$, $x \vee y \leq x \vee z$.*

Proof. We have

$$\begin{aligned} x \wedge y &= (x \wedge x) \wedge y && \text{(by (1.10))} \\ &= (x \wedge x) \wedge (y \wedge z) && \text{(by (1.15))} \\ &= (x \wedge y) \wedge (x \wedge z) && \text{(by (1.11))} \end{aligned}$$

Thus, by (1.14), $x \wedge y \leq x \wedge z$. The other part is an exercise. □

To emphasize the lattice operations $\leq, \wedge, \vee$, a lattice L is sometimes denoted by $\langle L, \leq, \wedge, \vee \rangle$.

Proposition 3 *In a lattice L, for any $x, y, z \in L$,*

$$x \wedge (y \vee z) = (x \wedge y) \vee (x \wedge z) \tag{1.15}$$

if, and only if

$$x \vee (y \wedge z) = (x \vee y) \wedge (x \vee z). \tag{1.16}$$

Proof. Let (1.15) hold, $x, y, z \in L$. Then

$$\begin{aligned}
x \vee (y \wedge z) &= x \vee (z \wedge y) && \text{(by (1.11))}\\
&= [x \vee (z \wedge x)] \vee (z \wedge y) && \text{(by (1.13))}\\
&= x \vee [(z \wedge x) \vee (z \wedge y)] && \text{(by (1.12))}\\
&= x \vee [z \wedge (x \vee y)] && \text{(by (1.15))}\\
&= [x \wedge (x \vee y)] \vee [z \wedge (x \vee y)] && \text{(by (1.13))}\\
&= [(x \vee y) \wedge x)] \vee [(x \vee y) \wedge z] && \text{(by (1.11))}\\
&= (x \vee y) \wedge (x \vee z) && \text{(by (1.15))}
\end{aligned}$$

Similarly, (1.16) implies (1.15) which completes the proof. □

Definition 2 *A lattice $\langle L, \leq, \wedge, \vee \rangle$ is called* distributive *if* (1.15), *or equivalently,* (1.16) *holds.*

Remark 4 *The family E of open sets of the real line as well as any chain are distributive lattices.*

Proof. Exercise.

Definition 3 *A distributive lattice $\langle L, \leq, \wedge, \vee \rangle$ in which with every element $x \in L$ there is associated an element $x^* \in L$ such that, for every $y \in L$,*

$$(x \wedge x^*) \vee y = y \text{ and } (x \vee x^*) \wedge y = y$$

is called a Boolean algebra. *The element x^* is called the* lattice-complement *of x,* l-complement *in short. A Boolean algebra is denoted by $\langle L, \leq, \wedge, \vee, ^* \rangle$.*

Example 8 Define on the set $\{0, 1\}$ $0^* = 1$ and $1^* = 0$. Then the system $\langle \{0, 1\}, \leq, \min, \max, ^* \rangle$ is a Boolean algebra. Define on the set $\mathcal{P}(A)$ of all subsets of a non-void set A, $X^* = A \setminus X$ for any subset $X \in \mathcal{P}(A)$. Then we obtain another Boolean algebra $\langle \mathcal{P}(A), \subseteq, \cap, \cup, ^* \rangle$.

Proposition 4 *Any complete chain L is* infinitely distributive, *i.e. for any $x \in L$ and any subset $\{y_i\}_{i \in \Gamma} \subseteq L$,*

$$x \wedge \bigvee_{i \in \Gamma} y_i = \bigvee_{i \in \Gamma} (x \wedge y_i), \tag{1.17}$$

$$x \vee \bigwedge_{i \in \Gamma} y_i = \bigwedge_{i \in \Gamma} (x \vee y_i), \tag{1.18}$$

where Γ is an index set.

Proof. To prove (1.17) we have two cases. (i) Assume $x \wedge y_i = y_i$ for each $i \in \Gamma$. Then $y_i \leq x$ for each $i \in \Gamma$ and $\bigvee_{i\in\Gamma}(x \wedge y_i) = \bigvee_{i\in\Gamma} y_i \leq x$. Therefore $x \wedge \bigvee_{i\in\Gamma} y_i = \bigvee_{i\in\Gamma} y_i = \bigvee_{i\in\Gamma}(x \wedge y_i)$. (ii) Let $x \wedge y_j = x$ for some $j \in \Gamma$. Then $x \leq y_j \leq \bigvee_{i\in\Gamma} y_i$ so that $x \wedge \bigvee_{i\in\Gamma} y_i = x = x \wedge y_j \leq \bigvee_{i\in\Gamma}(x \wedge y_i)$. On the other hand, for each $i \in \Gamma$ holds $x \wedge y_i \leq x \wedge \bigvee_{i\in\Gamma} y_i$, therefore $\bigvee_{i\in\Gamma}(x \wedge y_i) \leq x \wedge \bigvee_{i\in\Gamma} y_i$. We conclude that (1.17) is valid. (1.18) is left for the reader. □

Unlike the equations (1.15) and (1.16), the equations (1.17) and (1.18) are not, in general, equivalent with each other as we shall now see. Consider the distributive lattice $\langle E, \subseteq, \cap, \cup \rangle$ of open sets of the real line. For each family of open sets $X_i \in E, i \in \Gamma$, define $\bigvee_{i\in\Gamma} X_i = \bigcup_{i\in\Gamma} X_i$, and $\bigwedge_{i\in\Gamma} X_i$ is the greatest open set contained in every X_i. Let X be the open set $\mathcal{R} \setminus \{a\}$, where $a \in \mathcal{R}$. Define open sets $Y_i \in E$ by $Y_i = (a - 1/i, a + 1/i)$, $i \in \mathcal{N}$. Then $X \cap (\bigcup_{i\in\mathcal{N}} Y_i) = (a-1, a+1) \setminus \{a\} = \bigcup_{i\in\mathcal{N}}(X \cap Y_i)$. Thus (1.17) holds. On the other hand, $\bigwedge_{i\in\Gamma} Y_i = \emptyset$, thus $X \cup \bigwedge_{i\in\Gamma} Y_i = \mathcal{R} \setminus \{a\}$. Because $X \cup Y_i = \mathcal{R}$ for each $i \in \mathcal{N}$, also $\bigwedge_{i\in\mathcal{N}}(X \cup Y_i) = \mathcal{R}$. Therefore (1.18) does not hold.

1.2 Lattice Filters

A non-empty subset F of a lattice L is called a *lattice filter* of L or simply *filter* provided that, for all elements $a, b \in L$,

$$a \wedge b \in F \text{ iff } a, b \in F. \tag{1.19}$$

Remark 5 *A non-empty subset $F \subseteq L$ is a filter iff*

$$\text{if } a, b \in F, \text{ then } a \wedge b \in F, \tag{1.20}$$

$$\text{if } a \in F \text{ and } a \leq b, \text{ then } b \in F. \tag{1.21}$$

The condition (1.21) *can be replaced also by*

$$\text{if } a \in F \text{ and } b \in L, \text{ then } a \vee b \in F. \tag{1.22}$$

Proof. Exercise.

Of course, L itself is a filter of L. It is easy to see that given an element $a \in L$, the set $F = \{b \in L \mid a \leq b\}$ is a filter of L. If L contains the unit element $\mathbf{1}$, then also $\{\mathbf{1}\}$ is a filter of L. A filter F of L is called *proper* if $F \neq L$. If a lattice L contains the zero element $\mathbf{0}$ then, by (1.21), F is a proper filter iff $\mathbf{0} \notin F$. On the unit interval I, for any $a \in I$, $a \neq 0$, the subset $I_a = \{x \mid a \leq x \leq 1\}$ is a filter. By (1.21), if a lattice L contains the unit element $\mathbf{1}$, then $\mathbf{1}$ is included in every filter of L. The intersection $F \cap G$ of any two filters F, G of L, L containing the unit element $\mathbf{1}$, is again a filter of L. Indeed, $F \cap G \neq \emptyset$ as $\mathbf{1} \in F \cap G$ and $a, b \in F \cap G$ iff $a, b \in F, G$ iff $a \wedge b \in F, G$ iff $a \wedge b \in F \cap G$. Any non-empty family of filters of a lattice

L can be considered as a partial ordered set, the partial ordering relation being the set theoretical inclusion $\subseteq$. Thus, by a *chain of filters* of L we mean a non-empty family of filters such that, for any filters F, G in this family, either $F \subseteq G$ or $G \subseteq F$.

Proposition 5 *The union of any chain of proper filters of a lattice L containing the zero element $\mathbf{0}$ is a proper filter of L.*

Proof. Let $\mathcal{F} = \bigcup\{F_i \mid \cdots \subseteq F_k \subseteq F_l \subseteq \cdots \subseteq L\}$ be the union of a chain of proper filters of a lattice L, L containing $\mathbf{0}$. Then $a, b \in \mathcal{F}$ iff $a, b \in F_i$ for some index i iff $a \wedge b \in F_i \subseteq \mathcal{F}$. Hence $\mathcal{F}$ is a filter. Since no F_i contains $\mathbf{0}$, $\mathbf{0} \notin \mathcal{F}$. Thus, $\mathcal{F}$ is a proper filter of L. □

A proper filter F of L is called *maximal* if for any other filter G of L such that $F \subseteq G \subseteq L$, either $G = F$ or $G = L$. Thus, a maximal filter is a maximal element in the partial ordered set of all proper filters of L.

Proposition 6 *If a lattice L has the zero element then every proper filter F of L is contained in a maximal filter.*

Proof. Let $\mathcal{F}$ be a poset of all the proper filters of L. By Proposition 5, every chain of elements of $\mathcal{F}$ has an upper bound in $\mathcal{F}$. Thus, by Zorn's Lemma, for every filter $F \in \mathcal{F}$, there is a maximal element $G \in \mathcal{F}$, i.e. a maximal filter of L, such that $F \subseteq G$. □

Proposition 7 *If a lattice L has the zero element then, for every element $a \neq \mathbf{0}$ in L, there is a maximal filter F on L such that $a \in F$.*

Proof. $G = \{b \in L \mid a \leq b\}$ is a proper filter of L and is contained in a maximal filter F. Thus, $a \in F$. □

In particular, if L is a *non-degenerate* lattice, i.e. has at least two different elements and $\mathbf{0} \in L$ then, by taking $a \neq \mathbf{0}$ we have

Proposition 8 *Any non-degenerate lattice L containing $\mathbf{0}$ has a maximal filter.*

A proper filter F of L is called *prime* provided that, for any $a, b \in L$, $a \vee b \in F$ implies $a \in F$ or $b \in F$. On the unit interval I, for example, any proper filter I_a is a prime filter, the only maximal filter of I, however, is the set $\{x \mid 0 < x \leq 1\}$.

Proposition 9 *Each maximal filter F of a distributive lattice L is prime.*

Proof. Assume F is not prime. Then there exist $a, b \in L$ such that $a \vee b \in F$, $a \notin F$, $b \notin F$. Now the set $G = \{x \in L \mid a \wedge c \leq x, c \in F\}$ is a proper filter of L. Indeed, let $x, y \in G$. Then $a \wedge c \leq x$, $a \wedge d \leq y$, where $c, d \in F$.

Thus $a \wedge (c \wedge d) \leq x, y$ and therefore $a \wedge (c \wedge d) \leq x \wedge y$. Since F is a filter, $c \wedge d \in F$. Therefore $x \wedge y \in G$. Conversely, if $x \wedge y \in G$, then $a \wedge c \leq x \wedge y \leq x, y$ for some $c \in F$, hence $x, y \in G$. Assume $b \in G$. Then $a \wedge c \leq b$ for some $c \in F$. Since $b \leq a \vee b$, $a \vee b \in F$. Also $c \vee b \in F$, thus $b = (a \wedge c) \vee b = (a \vee b) \wedge (c \vee b) \in F$, a contradiction. Therefore $b \notin G$. This shows that G is a proper filter. Since $a \wedge c \leq c$ for any $c \in F$, we have $F \subseteq G$. Moreover, $a \notin F$, but because $a \wedge \mathbf{1} \leq a$, $\mathbf{1} \in F$, we have $a \in G$. Hence, F is not maximal, a contradiction. We conclude that F must be prime. □

1.3 Residuated Lattices

In the early studies of Fuzzy Set theory, the membership functions of union and intersection of two unit interval valued fuzzy sets were calculated by max- and min-operations, respectively, and the membership function of the complement of a fuzzy set was defined by means of the '1-' function. In Section 1.1 we saw that these operations are essentially lattice operations. In many applications, however, lattice structure alone is not rich enough to model fuzzy phenomena. In this Section we introduce an extremely important concept of residuated lattice. This structure appears, in one form or in an another, in practically all fuzzy inference systems, in the theory of fuzzy relations and, of course, in fuzzy logic.

A binary operation h defined on a non-void set A is called *isotone in the first variable* if, for any elements $a, b, c \in A$, $a \leq b$ implies $h(a, c) \leq h(b, c)$. h is called *antitone in the first variable* if $a \leq b$ implies $h(b, c) \leq h(a, c)$. The concepts *isotone in the second variable* and *antitone in the second variable* are defined in a similar way. If h is isotone in the first variable and isotone in the second variable h is called simply *isotone*. We are now ready to set an important

Definition 4 *Let $\langle L, \leq, \wedge, \vee\rangle$ be a lattice endowed by binary operations μ, f, g such that*

$$\mu \text{ is associative and isotone,} \tag{1.23}$$

$$\mu(x, y) \leq z \text{ iff } x \leq f(y, z) \text{ and } \mu(x, y) \leq z \text{ iff } y \leq g(x, z). \tag{1.24}$$

Then $\langle L, \leq, \wedge, \vee, \mu, f, g\rangle$ is a generalized residuated lattice.

We write xy instead of $\mu(x, y)$. The operations f, g are antitone in the first variable and isotone in the second variable, i.e. for any elements $x, y, z \in L$, if $x \leq y$ then

$$f(y, z) \leq f(x, z), g(y, z) \leq g(x, z), \tag{1.25}$$

$$f(z, x) \leq f(z, y), g(z, x) \leq g(z, y). \tag{1.26}$$

Indeed, let $x, y, z \in L$ and $x \leq y$. Since μ is isotone, we have $f(y,z)x \leq f(y,z)y$, where, by (1.24), $f(y,z)y \leq z$. Thus, $f(y,z)x \leq z$, which implies $f(y,z) \leq f(x,z)$. Similarly we verify that g is antitone in the first variable. By (1.24) and assumption, we have $f(z,x)z \leq x \leq y$, thus, by (1.24), $f(z,x) \leq f(z,y)$. Therefore (1.26) holds. The case for g is similar. Let μ be commutative. Then $g(y,z) \leq f(y,z)$ iff $g(y,z)y \leq z$ iff $yg(y,z) \leq z$ iff $g(y,z) \leq g(y,z)$. Similarly, $f(y,z) \leq g(y,z)$ iff $yf(y,z) \leq z$ iff $f(y,z)y \leq z$ iff $f(y,z) \leq f(y,z)$. Hence $f = g$. In such a case we write $x \to y$ instead of $f(x,y)$ or $g(x,y)$ and $x \odot y$ instead of xy. The couple $\langle \odot, \to \rangle$ is *adjoint couple*, $\odot$ is called *multiplication* or *product* and $\to$ is called *residuum*. Then (1.24) reads

$$x \odot y \leq z \text{ iff } x \leq y \to z. \tag{1.27}$$

The condition (1.27), called *Galois*[1] *correspondence*, will be extremely important. (1.27) implies e.g. that $x \leq y \to (x \odot y)$ and $(x \to y) \odot x \leq y$. These kind on (in-)equalities will be very frequent all over in this book. We will not cite (1.27) every time it is used.

Proposition 10 *Let $\langle \odot, \to \rangle$ be an adjoint couple. Then*

$$x \odot \bigvee_{i \in \Gamma} y_i = \bigvee_{i \in \Gamma} (x \odot y_i), \tag{1.28}$$

whenever these joins exist in L.

Proof. Let $x \in L$, $\{y_i\}_{i \in \Gamma} \subseteq L$. Assume both sides of the equation (1.28) exist in L. Since $\odot$ is isotone, $x \odot y_i \leq x \odot \bigvee_{i \in \Gamma} y_i$ for each $i \in \Gamma$, therefore

$$\bigvee_{i \in \Gamma} (x \odot y_i) \leq x \odot \bigvee_{i \in \Gamma} y_i.$$

Conversely, as $x \odot y_i \leq \bigvee_{i \in \Gamma} (x \odot y_i)$ for each $i \in \Gamma$ we have, by (1.27), $y_i \leq x \to \bigvee_{i \in \Gamma} (x \odot y_i)$ for each $i \in \Gamma$. Thus $\bigvee_{i \in \Gamma} y_i \leq x \to \bigvee_{i \in \Gamma} (x \odot y_i)$. Again by (1.27),

$$x \odot \bigvee_{i \in \Gamma} y_i \leq \bigvee_{i \in \Gamma} (x \odot y_i).$$

We conclude that (1.28) holds. □

Proposition 11 *Let L be a generalized residuated lattice. Assume the product is commutative. Then, for any $x, y \in L$,*

$$x \to y = \bigvee \{z \mid z \odot x \leq y\}. \tag{1.29}$$

Proof. Any $z \in L$ such that $z \odot x \leq y$ satisfies $z \leq x \to y$. On the other hand, $(x \to y) \odot x \leq y$. Therefore $(x \to y) \in \{z \mid z \odot x \leq y\}$. We conclude that (1.29) holds. □

[1] Evariste Galois 1811-32, French mathematician, romantic and revolutionary.

Proposition 12 *Let L be a complete lattice, and $\odot$ an isotone, associative and commutative binary operation on L such that* (1.28) *holds. Define another binary operation $\to$ on L by* (1.29). *Then $\langle \odot, \to \rangle$ is an adjoint couple. Moreover, $\to$ is uniquely defined.*

Proof. The first part is an exercise. The uniqueness of $\to$ follows immediately from (1.29). □

Definition 5 *If a generalized residuated lattice L with commutative product $\odot$ contains the elements $\mathbf{0}$, $\mathbf{1}$ and, for any $x \in L$, $x \odot \mathbf{1} = x$, then L is called a* residuated lattice.

A residuated lattice L is sometimes denoted by $\langle L, \leq, \wedge, \vee, \odot, \to, \mathbf{0}, \mathbf{1} \rangle$.

Example 1 Let p be a fixed natural number. Define on the real unit interval I binary operations $\odot$ and $\to$ by

$$x \odot y = 1 - \min\{1, \sqrt[p]{(1-x)^p + (1-y)^p}\}, \tag{1.30}$$

$$x \to y = \bigvee \{z \mid ; x \odot z \leq y\}. \tag{1.31}$$

Then $\langle I, \leq, \min, \max, \odot, \to, 0, 1 \rangle$, is a residuated lattice. Indeed, it is enough to show that the product defined by (1.30) is isotone, associative and commutative, and for each $x \in I$, $x \odot 1 = x$. This task we leave as an exercise for the reader.

Example 2 Let p be a fixed natural number. Define on the real unit interval I binary operations $\odot$ and $\to$ by

$$x \odot y = \sqrt[p]{\max\{0, x^p + y^p - 1\}}, \tag{1.32}$$

$$x \to y = \min\{1, \sqrt[p]{1 - x^p + y^p}\}. \tag{1.33}$$

Then $\langle I, \leq, \min, \max, \odot, \to, 0, 1 \rangle$ is a another residuated lattice, called *generalized Łukasiewicz structure*, in particular, *Łukasiewicz structure*[2] if $p = 1$. Indeed, it is easy to see that the operation $\odot$, defined by (1.32), is isotone, commutative and $x \odot 1 = x$ holds for any $x \in I$. $\odot$ is associative, too, as $(x \odot y) \odot z = 0 = x \odot (y \odot z)$ if $x^p + y^p + z^p \leq 2$ and $(x \odot y) \odot z = \sqrt[p]{x^p + y^p + z^p - 2} = x \odot (y \odot z)$ elsewhere. Moreover, for each $x, y, z \in I$, $x \odot y = \sqrt[p]{\max\{0, x^p + y^p - 1\}} \leq z$ iff $x^p + y^p - 1 \leq z^p$ iff $x^p \leq 1 - y^p + z^p$ iff $x \leq \min\{1, \sqrt[p]{1 - y^p + z^p}\}$ iff $x \leq y \to z$.

Example 3 Define on the real unit interval $x \odot y = \min\{x, y\}$, $x \to y = 1$ if $x \leq y$ and y otherwise. Then we obtain *Gödel structure*[3], a residuated lattice.

---

[2]This is the structure of set of values of truth in the infinite valued logic defined by Łukasiewicz in 1930's.

[3]According to Kurt Gödel. Sometimes this structure is called Brouwer structure, too, nominal to Brouwer who studied intuitionistic logic in the first half of the 20.th century and used similar structures in his studies.

Example 4 Let $\odot$ be the usual multiplication of real numbers on the unit interval and $x \to y = 1$ if $x \leq y$ and y/x otherwise. This structure is called *Products structure* (sometimes called also *Gaines structure*[4]), another residuated lattice.

Example 5 Any Boolean algebra can be regarded as a residuated lattice where the operations $\odot$ and $\wedge$ coincide and $x \to y = x^* \vee y$.

Example 6 Given two residuated lattices $\langle L_1, \leq, \wedge, \vee, \odot, \to, \mathbf{0}, \mathbf{1} \rangle$ and $\langle L_2, \leq, \wedge, \vee, \odot, \to, \mathbf{0}, \mathbf{1} \rangle$, the product set $L_1 \times L_2$ becomes a residuated lattice if one defines all the operations point wise, e.g.

$$(a, b) \odot (c, d) = (a \odot c, b \odot d) \text{ where } a, c \in L_1, b, d \in L_2.$$

This result holds for any finite or infinite product set of a collection of residuated lattices. Moreover, this construction is an example of a non-totally ordered residuated lattice. Indeed, in $L_1 \times L_2$ the order relation is defined by

$$(a, b) \leq (c, d) \text{ iff } a \leq c \text{ in } L_1, b \leq d \text{ in } L_2.$$

Since $\mathbf{0} \not\leq \mathbf{1}$ in all residuated lattices[5] we have that

$$(\mathbf{0}, \mathbf{1}) \not\leq (\mathbf{1}, \mathbf{0}) \text{ nor } (\mathbf{1}, \mathbf{0}) \not\leq (\mathbf{0}, \mathbf{1}).$$

Thus, $(\mathbf{0}, \mathbf{1})$ and $(1, 0)$ are incomparable elements in $L_1 \times L_2$.

Proposition 13 *In a residuated lattice L, for every $x, x_1, x_2, y, y_1, y_2 \in L$,*

$$x = \mathbf{1} \to x, \tag{1.34}$$
$$\mathbf{1} = x \to x, \tag{1.35}$$
$$x \odot y \leq x, y, \tag{1.36}$$
$$x \odot y \leq x \wedge y, \tag{1.37}$$
$$y \leq x \to y, \tag{1.38}$$
$$x \odot y \leq x \to y, \tag{1.39}$$
$$x \leq y \text{ iff } \mathbf{1} = x \to y, \tag{1.40}$$
$$\mathbf{1} = x \to y = y \to x \text{ iff } x = y, \tag{1.41}$$
$$x \to \mathbf{1} = \mathbf{1}, \tag{1.42}$$
$$0 \to x = \mathbf{1}, \tag{1.43}$$
$$x \to (y \to x) = \mathbf{1}, \tag{1.44}$$
$$(x \to y) \to [(y \to z) \to (x \to z)] = \mathbf{1}, \tag{1.45}$$

[4]Gaines studied this structure in fuzzy logic framework in 1976.
[5]Containing at least two elements!

$$(x \to y) \to [(z \to x) \to (z \to y)] = \mathbf{1}, \quad (1.46)$$

$$(x \odot y) \to z = x \to (y \to z), \quad (1.47)$$

$$x \to (y \to z) = y \to (x \to z), \quad (1.48)$$

$$(x_1 \to y_1) \to \{(y_2 \to x_2) \to [(y_1 \to y_2) \to (x_1 \to x_2)]\} = \mathbf{1}. \quad (1.49)$$

Proof. To verify (1.47), we reason that $[x \to (y \to z)] \odot (x \odot y) \leq (y \to z) \odot y \leq z$. Thus $x \to (y \to z) \leq (x \odot y) \to z$. Conversely, $(x \odot y) \to z \leq x \to (y \to z)$ iff $[(x \odot y) \to z] \odot x \leq y \to z$ iff $[(x \odot y) \to z] \odot (x \odot y) \leq z$ iff $(x \odot y) \to z \leq (x \odot y) \to z$. Hence (1.47) holds. By using several times (1.27), we realize that (1.49) holds iff

$$(x_1 \to y_1) \odot (y_2 \to x_2) \odot (y_1 \to y_2) \odot x_1 \leq x_2,$$

which, indeed, is the case as

$$\begin{aligned}(x_1 \to y_1) \odot (y_2 \to x_2) \odot (y_1 \to y_2) \odot x_1 &\leq (y_2 \to x_2) \odot (y_1 \to y_2) \odot y_1 \\ &\leq (y_2 \to x_2) \odot y_2 \\ &\leq x_2.\end{aligned}$$

The remaining part of the proof is left for the reader. □

Later, when we talk about fuzzy logic, the residuum operation $\to$ will play an essential role as it offers a nice way to interpret fuzzy logic *implication*, as our notation already suggests. In Exercises, we introduce some more properties of residuated lattices. In a residuated lattice L, we introduce an abbreviation $x^* = x \to \mathbf{0}$, called *complement* of an element $x \in L$. This abbreviation will turn out to be important in fuzzy logic when we discuss about *negation.* We write x^{**} instead of $(x^*)^*$. In general, the complement x^* of an element x in a residuated lattice L does not coincide with the l-complement of x if the later exists. In Boolean algebras, however, complement and l-complement coincides, as we shall later see.

Proposition 14 *In a residuated lattice L, for every $x, y \in L$,*

$$x \odot x^* = \mathbf{0}, \quad (1.50)$$

$$x \leq x^{**}, \quad (1.51)$$

$$\mathbf{1}^* = \mathbf{0}, \quad (1.52)$$

$$\mathbf{0}^* = \mathbf{1}, \quad (1.53)$$

$$x \to y \leq y^* \to x^*, \quad (1.54)$$

$$x^* = x^{***}. \quad (1.55)$$

Proof. By (1.27), $x \leq x^{**}$ iff $x \odot x^* \leq \mathbf{0}$ iff $x^* \leq x \to \mathbf{0}$. Thus, (1.50) and (1.51) hold. $\mathbf{1}^* \leq \mathbf{0}$ iff $\mathbf{0} = \mathbf{1} \to \mathbf{0}$ iff $\mathbf{0} \odot \mathbf{1} \leq \mathbf{0}$, which is the case, therefore also $\mathbf{1} \leq \mathbf{0}^*$. We conclude that (1.52) and (1.53) hold. Take $z = \mathbf{0}$ in (1.45), then we have $\mathbf{1} = (x \to y) \to (x^* \to y^*)$, hence $x \to y \leq x^* \to y^*$, i.e. (1.54) holds. To verify (1.55) we reason that $x^* \leq x^{***}$ holds by (1.51).

Since $x \leq x^{**} \leq x^{****}$, we have that $x \odot x^{***} \leq \mathbf{0}$. Therefore $x^{***} \leq x^*$. We conclude that (1.55) is valid. The proof is complete. □

By *bi-residuum* on a residuated lattice L we shall understand the derived operation $x \leftrightarrow y = (x \to y) \wedge (y \to x)$. Bi-residuum will offer us an elegant way to interpret fuzzy logic *equivalence* and *fuzzy similarity relation.*

Proposition 15 *Bi-residuum has the following properties*

$$x \leftrightarrow \mathbf{1} = x, \tag{1.56}$$

$$x = y \text{ iff } x \leftrightarrow y = \mathbf{1}, \tag{1.57}$$

$$x \leftrightarrow y = y \leftrightarrow x, \tag{1.58}$$

$$(x \leftrightarrow y) \odot (y \leftrightarrow z) \leq x \leftrightarrow z, \tag{1.59}$$

$$(x_1 \leftrightarrow y_1) \wedge (x_2 \leftrightarrow y_2) \leq (x_1 \wedge x_2) \leftrightarrow (y_1 \wedge y_2), \tag{1.60}$$

$$(x_1 \leftrightarrow y_1) \wedge (x_2 \leftrightarrow y_2) \leq (x_1 \vee x_2) \leftrightarrow (y_1 \vee y_2), \tag{1.61}$$

$$(x_1 \leftrightarrow y_1) \odot (x_2 \leftrightarrow y_2) \leq (x_1 \odot x_2) \leftrightarrow (y_1 \odot y_2), \tag{1.62}$$

$$(x_1 \leftrightarrow y_1) \odot (x_2 \leftrightarrow y_2) \leq (x_1 \to x_2) \leftrightarrow (y_1 \to y_2). \tag{1.63}$$

Proof. (1.56), (1.57) and (1.58) are left for the reader as an exercise. By (1.45) and (1.27), $(x \to y) \odot (y \to z) \leq x \to z$, therefore $(x \leftrightarrow y) \odot (y \leftrightarrow z) \leq (x \to y) \odot (y \to z) \leq x \to z$. Similarly, $(x \leftrightarrow y) \odot (y \leftrightarrow z) \leq z \to x$. We conclude that (1.59) holds. Denote $A = (x_1 \leftrightarrow y_1)$, $B = (x_2 \leftrightarrow y_2)$. Now,

$$\begin{aligned} [A \wedge B] \odot (x_1 \wedge x_2) &\leq [(x_1 \to y_1) \wedge (x_2 \to y_2)] \odot (x_1 \wedge x_2) \\ &\leq [(x_1 \to y_1) \odot x_1] \wedge [(x_2 \to y_2) \odot x_2] \\ &\leq y_1 \wedge y_2. \end{aligned}$$

Therefore $(x_1 \leftrightarrow y_1) \wedge (x_2 \leftrightarrow y_2) \leq (x_1 \wedge x_2) \to (y_1 \wedge y_2)$. By a similar argument we see that also $(x_1 \leftrightarrow y_1) \wedge (x_2 \leftrightarrow y_2) \leq (y_1 \wedge y_2) \to (x_1 \wedge x_2)$ holds. We conclude (1.60). Moreover,

$$\begin{aligned} [A \wedge B] \odot (x_1 \vee x_2) &= \{[A \wedge B] \odot x_1\} \vee \{[A \wedge B] \odot x_2\} \\ &\leq [(x_1 \to y_1) \odot x_1] \vee [(x_2 \to y_2) \odot x_2] \\ &\leq y_1 \vee y_2, \end{aligned}$$

$$\begin{aligned} A \odot B \odot (x_1 \odot x_2) &\leq [(x_1 \to y_1) \odot x_1] \odot [(x_2 \to y_2) \odot x_2] \\ &\leq y_1 \odot y_2, \end{aligned}$$

$$\begin{aligned} A \odot B \odot (x_1 \to x_2) &\leq (y_1 \to x_1) \odot (x_2 \to y_2) \odot (x_1 \to x_2) \\ &\leq (y_1 \to x_2) \odot (x_2 \to y_2) \\ &\leq y_1 \to y_2. \end{aligned}$$

(1.61), (1.62) and (1.63) now follow by the Galois correspondence (1.27) from these three (in-)equalities, respectively. □

At the end of this Section we focus on complete residuated lattices. We have already seen that multiplication preserves joins, i.e. that equation (1.28) holds. In particular, as $\langle \wedge, \Rightarrow \rangle$ can be regarded as an adjoint couple, where the residuum $\Rightarrow$ is defined via (1.29), we conclude that (1.18) holds in any complete residuated lattice.

Proposition 16 *In a complete residuated lattice L, for every element $x \in L$ and every subset $\{y_i\}_{i \in \Gamma} \subseteq L$,*

$$x \to \bigwedge_{i \in \Gamma} y_i = \bigwedge_{i \in \Gamma} (x \to y_i), \tag{1.64}$$

$$\bigvee_{i \in \Gamma} y_i \to x = \bigwedge_{i \in \Gamma} (y_i \to x), \tag{1.65}$$

$$\bigvee_{i \in \Gamma} (y_i \to x) \leq \bigwedge_{i \in \Gamma} y_i \to x, \tag{1.66}$$

$$\bigvee_{i \in \Gamma} (x \to y_i) \leq x \to \bigvee_{i \in \Gamma} y_i. \tag{1.67}$$

Proof. Since $\to$ is antitone in the first variable, for every $i \in \Gamma$ holds $y_i \to x \leq \bigwedge_{i \in \Gamma} y_i \to x$, thus $\bigvee_{i \in \Gamma}(y_i \to x) \leq \bigwedge_{i \in \Gamma} y_i \to x$, hence (1.66) is valid. Similarly, since $\to$ is isotone in the second variable, (1.67) holds.(1.64), (1.65) are left for the reader as an exercise. □

The (in-)equations (1.64)-(1.67) hold, of course, in incomplete residuated lattices, too, whenever they make sense. In particular, by taking $x = \mathbf{0}$ in (1.65) and (1.66), we see that

$$(\bigvee_{i \in \Gamma} y_i)^* = \bigwedge_{i \in \Gamma} y_i^*, \tag{1.68}$$

$$\bigvee_{i \in \Gamma} y_i^* \leq (\bigwedge_{i \in \Gamma} y_i)^*. \tag{1.69}$$

We abbreviate $a^n = \overbrace{a \odot \ldots \odot a}^{n \text{ terms}}$, in particular, a^0 stands for $\mathbf{1}$.

Proposition 17 *Let L be a residuated lattice. For every $a, b_1, b_2, c_1, c_2 \in L$, if $a \leq b_1 \leftrightarrow b_2$, $a \leq c_1 \leftrightarrow c_2$, then $a^2 \leq (b_1 \to c_1) \leftrightarrow (b_2 \to c_2)$.*

Proof. The condition $a \leq b_1 \leftrightarrow b_2$ implies $a \leq b_2 \to b_1$, or equally, $a \odot b_2 \leq b_1$ and similarly, $a \odot c_1 \leq c_2$. Now $a \odot a \leq (b_1 \to c_1) \to (b_2 \to c_2)$ iff $a \odot a \odot (b_1 \to c_1) \odot b_2 \leq c_2$, which is the case as $a \odot a \odot (b_1 \to c_1) \odot b_2 \leq a \odot (b_1 \to c_1) \odot b_1 \leq a \odot c_1 \leq c_2$. Similarly, $a \odot a \leq (b_2 \to c_2) \to (b_1 \to c_1)$. Hence, $a^2 \leq (b_1 \to c_1) \leftrightarrow (b_2 \to c_2)$. □

Proposition 18 *In a complete residuated lattice L, for every $a, b_i, c_i \in L$, $i \in \Gamma$, if $a \leq b_i \leftrightarrow c_i$, then $a \leq (\bigwedge_{i\in\Gamma} b_i) \leftrightarrow (\bigwedge_{i\in\Gamma} c_i)$.*

Proof. $a \leq b_i \leftrightarrow c_i$ for each $i \in \Gamma$ implies $a \odot \bigwedge_{i\in\Gamma} b_i \leq a \odot b_i \leq c_i$. Thus $a \odot \bigwedge_{i\in\Gamma} b_i \leq \bigwedge_{i\in\Gamma} c_i$. Therefore $a \leq (\bigwedge_{i\in\Gamma} b_i) \rightarrow (\bigwedge_{i\in\Gamma} c_i)$. Similarly $a \leq (\bigwedge_{i\in\Gamma} c_i) \rightarrow (\bigwedge_{i\in\Gamma} b_i)$. Hence $a \leq (\bigwedge_{i\in\Gamma} b_i) \leftrightarrow (\bigwedge_{i\in\Gamma} c_i)$. □

Proposition 19 *In a residuated lattice L, for every $a_i, b_i, b_{i+1} \in L$, $i = 1 \dots n$, if $a_i \leq b_i \leftrightarrow b_{i+1}$ then $a_1 \odot \dots \odot a_n \leq b_1 \leftrightarrow b_{n+1}$.*

Proof. Exercise.

1.4 BL-Algebras

Definition 6 *A residuated lattice $\langle L, \leq, \wedge, \vee, \odot, \rightarrow, \mathbf{0}, \mathbf{1}\rangle$ is called a* BL-algebra, *if the following three identities hold*[6] *for all $x, y \in L$*

$$x \wedge y = x \odot (x \rightarrow y), \tag{1.70}$$

$$x \vee y = [(x \rightarrow y) \rightarrow y] \wedge [(y \rightarrow x) \rightarrow x], \tag{1.71}$$

$$(x \rightarrow y) \vee (y \rightarrow x) = \mathbf{1}. \tag{1.72}$$

Proposition 20 *A linear residuated lattice is a BL-algebra iff the equation (1.70) holds in it.*

Proof. We have to show that in a linear residuated lattice L, (1.70) implies (1.71) and (1.72). Since, for all $x, y \in L$, $x \leq y$ or $y \leq x$, we have $x \rightarrow y = \mathbf{1}$ or $y \rightarrow x = \mathbf{1}$, hence $(x \rightarrow y) \vee (y \rightarrow x) = \mathbf{1}$, whence (1.72) holds. Assume $x \leq y$. Then $y = x \vee y$ and $(x \rightarrow y) \rightarrow y = \mathbf{1} \rightarrow y = y$ and, moreover, since $(y \rightarrow x) \odot y \leq x$, we have $y \leq (y \rightarrow x) \rightarrow x$. Therefore $x \vee y = [(x \rightarrow y) \rightarrow y] \wedge [(y \rightarrow x) \rightarrow x]$. The case for $y \leq x$ is symmetric. Thus (1.71) holds.□

Lukasiewicz structure, Gödel structure and Product structure are BL-algebras as will be shown in Exercise 1. Not every residuated lattice, however, is a BL-algebra. Consider, for example, a residuated lattice defined on the unit interval, for all $x, y, z \in [0, 1]$, such that

$$x \odot y = \begin{cases} 0 & \text{if } x + y \leq \frac{1}{2} \\ x \wedge y & \text{elsewhere} \end{cases}, \quad x \rightarrow y = \begin{cases} 1 & \text{if } x \leq y \\ \max\{\frac{1}{2} - x, y\} & \text{elsewhere.} \end{cases}$$

Let $0 < y < x$, $x + y < \frac{1}{2}$. Then $y < \frac{1}{2} - x$ and $0 \neq y = x \wedge y$, but $x \odot (x \rightarrow y) = x \odot (\frac{1}{2} - x) = 0$. Therefore (1.70) does not hold.□

Proposition 21 *BL-algebras are distributive lattices.*

[6] This definition is due to Hájek. The letters BL stand for basic (fuzzy) logic

Proof. Let x, y, z be elements of a BL-algebra L. First we realize that $(x \wedge y), (x \wedge z) \leq x \wedge (y \vee z)$, therefore $(x \wedge y) \vee (x \wedge z) \leq x \wedge (y \vee z)$. The converse holds, too, as by (1.70), (1.28) and since $\rightarrow$ is antitone in the first variable, we have

$$\begin{aligned} x \wedge (y \vee z) &= (y \vee z) \odot [(y \vee z) \rightarrow x] \\ &= \{y \odot [(y \vee z) \rightarrow x]\} \vee \{z \odot [(y \vee z) \rightarrow x]\} \\ &\leq [y \odot (y \rightarrow x)] \vee [z \odot (z \rightarrow x)] \\ &= (y \wedge x) \vee (z \wedge x) \\ &= (x \wedge y) \vee (x \wedge z). \end{aligned}$$

thus (1.15) holds. □

Definition 7 *Let L be a BL-algebra. A subset $D \subseteq L$ is a* deductive system *of L, ds for short, if the following conditions are satisfied*

$$\mathbf{1} \in D, \tag{1.73}$$

$$\textit{if } x, x \rightarrow y \in D \textit{ then } y \in D. \tag{1.74}$$

Clearly, L and $\{\mathbf{1}\}$ are deductive systems of L. If $D \subseteq L$ is a *ds*, $a \in D$, $a \leq b$, then $a \rightarrow b = \mathbf{1}$, thus $b \in D$. A deductive system D is called *proper* if there is an element $a \in L$ such that $a \notin D$. We immediately have

Remark 6 *A deductive system D is proper iff $\mathbf{0} \notin \mathbf{D}$ iff for no element $a \in L$ holds $a, a^* \in D$.*

In Exercise 7 we have another characterization for a deductive system.

Proposition 22 *A deductive system $D \subseteq L$ is a lattice filter of L.*

Proof. Let $a, b \in D$. Since $a \rightarrow [b \rightarrow (a \odot b)] = \mathbf{1} \in D$, we have $b \rightarrow (a \odot b) \in D$, therefore $a \odot b \in D$. Now $a \odot b \leq a \odot \mathbf{1} = a$, $a \odot b \leq \mathbf{1} \odot b = b$, hence $a \odot b \leq a \wedge b$ and we conclude $a \wedge b \in D$. Conversely, if $a \wedge b \in D$, then also $a, b \in D$ as $a \wedge b \leq a, b$. Thus D is a lattice filter of L. □

Since $a \rightarrow (a \rightarrow \ldots (a^n) \ldots) = \mathbf{1}$, we reason that if a is in a deductive system D, then also $a^n \in D$. Because $a^n \leq a$, the converse holds, too. Thus, we have

Remark 7 *Let D be a deductive system of a BL-algebra L, $a \in L$. Then $a \in D$ iff $a^n \in D$ for any natural number n.*

Deductive systems are called also *implicative filters* in literature. To avoid confusion we reserve, however, the name filter to lattice filters in this book. Filters are not, in general, deductive systems. In Łukasiewicz structure, for example, the only proper *ds* is $\{1\}$. To see this we realize first that, for any $a \in I$, $a^n = \max\{0, na - (n-1)\}$. Assume now $a \neq 1$ is in some deductive system D. Then there is a natural number m such that $a \leq \frac{m-1}{m}$. Since $a^m \in D$ and $a^m = 0$, we conclude that D is not proper. We have, however,

Proposition 23 *In Boolean algebras as well as in Brouwer structure deductive systems and filters coincide.*

Proof. Let F be a filter of a Boolean algebra L. Trivially $\mathbf{1} \in F$. Assume $a, a \to b \in F$. Then $a \wedge (a \to b) \in F$. Since $a \wedge (a \to b) \leq b$, we have $b \in F$. Therefore F is a *ds*. The converse holds by Proposition 22. The same argument holds for Brouwer structure, too.□

Proposition 24 *Let X, Y be two deductive systems of a BL-algebra L. Define a set*

$$D_{X \cup Y} = \{a \in L \mid y_1 \odot \ldots \odot y_m \leq a \text{ for some } y_1, \ldots, y_m \in X \cup Y\}.$$

Then $D_{X \cup Y}$ is a ds of L and $X \cup Y \subseteq D_{X \cup Y}$.

Proof. Trivially $\mathbf{1} \in D_{X \cup Y}$. Let $a, a \to b \in D_{X \cup Y}$. Then $y_1 \odot \ldots \odot y_m \leq a$ for some $y_1, \ldots, y_m \in X \cup Y$, $z_1 \odot \ldots \odot z_p \leq a \to b$ for some $z_1, \ldots, z_p \in X \cup Y$. Since $z_1 \odot \ldots \odot z_p \odot a \leq b$, we conclude that $z_1 \odot \ldots \odot z_p \odot y_1 \odot \ldots \odot y_m \leq b$. Therefore $b \in D_{X \cup Y}$. Thus, $D_{X \cup Y}$ is a *ds*. Moreover, as any $y \in X \cup Y$, $y \leq y$ we have $X \cup Y \subseteq D_{X \cup Y}$. □

A proper deductive system D of L is called *prime* if, for any $a, b \in L$, the condition $a \vee b \in D$ implies $a \in D$ or $b \in D$. We leave for the reader as an exercise to prove

Remark 8 *If D is a prime ds then, for any $a, b \in L$, either $a \to b \in D$ or $b \to a \in D$. Moreover, L is linear iff any proper ds of L is prime.*

An BL-algebra L is called *degenerate* if $\mathbf{0} = \mathbf{1}$ in L. Otherwise L is *non-degenerate.*

Theorem 1 *A non-degenerate BL-algebra contains a prime ds.*

Proof. Let L be a non-degenerate BL-algebra. By Proposition 8, L contains a maximal filter which, by Proposition 9, is a prime filter. Denote it by P and $P^c = L \setminus P$. Then $P^c \neq \emptyset$. Define

$$\hat{\mathrm{P}} = \bigcap_{y \in P^c} \{x \mid x \to y \in P^c\}.$$

We show that $\hat{\mathrm{P}}$ is a prime *ds*. Since $\mathbf{1} \to y = y$ for any $y \in P^c$, we conclude that $\mathbf{1} \in \hat{P}$. Assume $x, x \to z \in \hat{\mathrm{P}}$. Then $x \to y \in P^c$ for any $y \in P^c$, $(x \to z) \to w \in P^c$ for any $w \in P^c$, in particular, $(x \to z) \to (x \to y) \in P^c$ for any $y \in P^c$. Since $z \to y \leq (x \to z) \to (x \to y)$, assumption $z \to y \in P$ would imply $(x \to z) \to (x \to y) \in P$, a contradiction. Therefore $z \to y \in P^c$ for any $y \in P^c$, which in turn implies $z \in \hat{\mathrm{P}}$. Thus, $\hat{\mathrm{P}}$ is a *ds*. $\hat{\mathrm{P}}$ is proper as, for any $y \in P^c$, $y \to y = \mathbf{1} \in P$. Thus $y \notin \hat{\mathrm{P}}$. The same argument shows that $\hat{\mathrm{P}} \subseteq P$. Finally, to prove that $\hat{\mathrm{P}}$ is prime, assume $x, y \in L$, $x \vee y \in \hat{\mathrm{P}}$, but $x \notin \hat{\mathrm{P}}$, $y \notin \hat{\mathrm{P}}$. Then there exist $z, w \in P^c$ such that

$x \to z, y \to w \in P$. Necessarily $z \vee w \in P^c$ as P is a prime filter. Therefore $(x \vee y) \to (z \vee w) \in P^c$. On the other hand, since $x \to z \leq x \to (z \vee w)$, $y \to w \leq y \to (z \vee w)$, we conclude that $x \to (z \vee w), y \to (z \vee w) \in P$ and, by (1.65), $[x \to (z \vee w)] \wedge [y \to (z \vee w)] = (x \vee y) \to (z \vee w) \in P$, a contradiction. Thus $x \in \hat{\mathrm{P}}$ or $y \in \hat{\mathrm{P}}$. We have seen that $\hat{\mathrm{P}}$ is a prime *ds*. □

Proposition 25 *Let P be a prime ds of a BL-algebra L. If D is a proper ds such that $P \subseteq D$, then also D is prime.*

Proof. Assume $a, b \in L$ such that $a \vee b \in D$. Since P is prime either $a \to b \in P$ or $b \to a \in P$. Assume $a \to b \in P$. Then $a \to b \in D$. Since $a \vee b \leq (a \to b) \to b \in D$, we conclude that $b \in D$. In a similar manner the condition $b \to a \in P$ implies $a \in D$. Thus, D is prime. □

We leave for the reader as an exercise to verify the following

Remark 9 *The set of proper deductive systems containing a given prime ds is totally ordered with respect to the set theoretical inclusion.*

If D is a proper *ds* and there exists another proper *ds* E such that $D \subseteq E$ we say that D can be *extended* to E.

Theorem 2 *Any proper ds F of a non-degenerate BL-algebra L can be extended to a prime ds.*

Proof. Let F be a proper *ds*. By Proposition 22, F is a filter on L. By Proposition 6, F is contained in a maximal filter P which, by Proposition 9, is a prime filter. By Theorem 1, P defines a prime *ds* $\hat{\mathrm{P}}$ such that $\hat{\mathrm{P}} \subseteq P$. Since F and $\hat{\mathrm{P}}$ are two proper *ds*, the set $D_{F \cup \hat{\mathrm{P}}}$ is a *ds* and $F \subseteq D_{F \cup \hat{\mathrm{P}}}$. We demonstrate that $D_{F \cup \hat{\mathrm{P}}}$ is proper by showing that $D_{F \cup \hat{\mathrm{P}}} \subseteq P$. Let $x \in D_{F \cup \hat{\mathrm{P}}}$. Then $z_1 \odot \ldots \odot z_n \odot y_1 \odot \ldots \odot y_m \leq x$, where, by the commutativity of $\odot$, we may assume that $z_1 \ldots z_n \in F$ and $y_1 \ldots y_m \in \hat{\mathrm{P}}$. Since

$$z_1 \to (\ldots \to (z_n \to (y_1 \to (\ldots \to (y_m \to x) \ldots))) \ldots) = \mathbf{1} \in F,$$

we conclude that $y_1 \to (\ldots \to (y_m \to x) \ldots) \in F \subseteq P$. Assume now $x \in P^c$. Then, by definition of $\hat{\mathrm{P}}$, $y_m \to x \in P^c$, therefore $y_{m-1} \to (y_m \to x) \in P^c$, etc. and finally $y_1 \to (\ldots \to (y_m \to x) \ldots) \in P^c$, a contradiction. Thus $x \in P$. Therefore $D_{F \cup \hat{\mathrm{P}}}$ is proper. Since $\hat{\mathrm{P}} \subseteq D_{F \cup \hat{\mathrm{P}}}$ and $\hat{\mathrm{P}}$ is prime, $D_{F \cup \hat{\mathrm{P}}}$ is prime by Proposition 25. □

A proper *ds* D of a BL-algebra L is called *maximal*, if it is a maximal element in the partially ordered set of all proper *ds* of L, i.e. for any other proper *ds* E of L such that $D \subseteq E$ holds $E = D$.

Theorem 3 *In a non-degenerate BL-algebra L, any proper ds can be extended to a maximal, prime ds.*

Proof. Let $D \subseteq L$ be a proper *ds*. By Theorem 2, D can be extended to a prime *ds* E. Set

$$\mathcal{F} = \{G \mid E \subseteq G, G \text{ a proper } ds \text{ on } L\}.$$

By Remark 9, $\mathcal{F}$ is a totally ordered set and, by Proposition 25, any $G \in \mathcal{F}$ is a prime *ds*. Define

$$M = \bigcup\{G \mid G \in \mathcal{F}\}.$$

Then trivially $\mathbf{1} \in M$. If $a, a \to b \in M$, then $a, a \to b \in G$ for some $G \in \mathcal{F}$. Thus $b \in G \subseteq M$. Therefore M is a *ds*. Since $\mathbf{0} \notin G$ for any $G \in \mathcal{F}$, $\mathbf{0} \notin M$. Thus M is proper and obviously prime. The maximality of M is implied by the construction of M. The proof is complete. □

If $\sim$ is a congruence relation on a BL-algebra L, that is, an equivalence relation which is a congruence with respect to the operations $\to$, $\odot$, *, $\vee$ and $\wedge$, then $\sim$ is called *maximal* if $\sim$ is not the trivial equivalence relation (i.e. there are $x, y \in L$ such that $x \not\sim y$) and whenever $x \equiv y$, where $\equiv$ is any non-trivial congruence on L, then also $x \sim y$. Now we show that there is one-to-one correspondence between (maximal) deductive systems of L and (maximal) congruence relations on L. The verification of the following proposition, however, is left to the reader.

Proposition 26 *If $\sim$ is a congruence relation on a BL-algebra L then $D = \{a \in L \mid a \sim \mathbf{1}\}$ is a (maximal) deductive system of L.*

We also have

Proposition 27 *If D is a deductive system of L then, by defining $x \sim y$ iff $(x \to y) \odot (y \to x) \in D$, we obtain a congruence relation on L.*

Proof. First we show that $\sim$ is an equivalence relation. Let $x, y, z \in L$. Trivially $x \sim x$ and, if $x \sim y$ then $y \sim x$. To show that $\sim$ is transitivity, let $x \sim y$, $y \sim z$. Then $(x \to y) \odot (y \to x) \leq x \to y$ and $(y \to z) \odot (z \to y) \leq y \to z$. Thus $x \to y \in D$, $y \to z \in D$, hence $(x \to y) \odot (y \to z) \in D$. Since $(x \to y) \odot (y \to z) \leq x \to z$ we conclude $x \to z \in D$. By a similar argument $z \to x \in D$. Therefore $(x \to z) \odot (z \to x) \in D$. Thus $x \sim z$. Then we show that $\sim$ is a congruence relation with respect to $\to$. Let $x \sim y$, $z \sim w$. Then $x \to y \in D$, $w \to z \in D$. By (1.49), $(x \to y) \to \{(w \to z) \to [(y \to w) \to (x \to z)]\} = \mathbf{1} \in D$. Therefore $(w \to z) \to [(y \to w) \to (x \to z)] \in D$ and so $[(y \to w) \to (x \to z)] \in D$. By a symmetric argument $[(x \to z) \to (y \to w)] \in D$. Therefore

$$[(x \to z) \to (y \to w)] \odot [(y \to w) \to (x \to z)] \in D,$$

that is $(x \to z) \sim (y \to w)$. In particular, if $z = w = \mathbf{0}$, we realize $x^* \sim y^*$. To see that $\sim$ is a congruence relation with respect to $\odot$ we assume first that $x \sim y$, $z \in L$. Since $y \leq z \to (y \odot z)$, we have $x \to y \leq x \to [z \to (y \odot z)] =$

$(x \odot z) \to (y \odot z)$. But $x \to y \in D$, so is $(x \odot z) \to (y \odot z) \in D$. By symmetry, $(y \odot z) \to (x \odot z) \in D$. Therefore $[(y \odot z) \to (x \odot z)] \odot [(x \odot z) \to (y \odot z)] \in D$, and so $(x \odot z) \sim (y \odot z)$. Trivially $(y \odot z) \sim (z \odot y)$. Assume now $x \sim y$ and $z \sim w$. Then we have a desired result $(x \odot z) \sim (y \odot z) \sim (z \odot y) \sim (w \odot y)$. We leave as an exercise for the reader to verify that $\sim$ is a congruence with respect to the lattice operations $\wedge$ and $\vee$, too.□

Proposition 28 *In a BL-algebra L, there is one-to-one correspondence between (maximal) deductive systems of L and (maximal) congruence relations on L.*

Proof. Let a *ds* D be given and define $x \sim_D y$ iff $(x \to y) \odot (y \to z) \in D$. Then $F = \{a \in L \mid a \sim_D \mathbf{1}\}$ is a *ds* and $F = D$. Indeed, if $a \in F$ then $a = (a \to \mathbf{1}) \odot (\mathbf{1} \to a) \in D$ and, if $a \in D$, then $a = (a \to \mathbf{1}) \odot (\mathbf{1} \to a) \in D$, thus $a \sim_D \mathbf{1}$, hence $a \in F$. Assume a congruence relation $\sim$ on L is given and define a set $D = \{a \in L \mid a \sim \mathbf{1}\}$. Then the relation $x \sim_D y$ iff $(x \to y) \odot (y \to z) \in D$ is a congruence and coincides with $\sim$. Indeed, if $x \sim_D y$ then $(x \to y) \odot (y \to z) \in D$, thus $(x \to y) \in D$, hence $(x \to y) \sim \mathbf{1}$. Therefore $[x \odot (x \to y)] \sim [x \odot \mathbf{1}] = x$, hence $(x \wedge y) \sim x$. By a similar argument $(y \wedge x) \sim y$. Since $(x \wedge y) \sim (y \wedge x)$, we have $x \sim y$. Conversely, if $x \sim y$, then $(x \to y) \sim (y \to y) = \mathbf{1}$ and $(y \to x) \sim (x \to x) = \mathbf{1}$. Thus, $[(x \to y) \odot (y \to x)] \sim [\mathbf{1} \odot \mathbf{1}] = \mathbf{1}$, whence $[(x \to y) \odot (y \to x)] \in D$ and, therefore $x \sim_D y$. We conclude $x \sim y$ iff $x \sim_D y$. The verification of D is maximal iff $\sim_D$ is maximal is left for the reader.□

Given a deductive system D, we denote by L/D the induced set of equivalent classes $\{|x| \mid x \in L\}$. By definition, $|x| = |\mathbf{1}|$ iff $x \in D$. Our aim now is to show that the *quotient algebra* L/D can be regarded as a BL-algebra. Indeed, we have

Proposition 29 *Let D be a ds of a BL-algebra L. Define on L/D, for all* $x, y \in L$, $|x| \leq |y|$ *iff* $x \to y \in D$. *Then* $\leq$ *is an order relation on L/D and* $|\mathbf{0}|, |\mathbf{1}|$ *are the least element and the largest element, respectively.*

Proof. Since $x \to x \in D$, we have $|x| \leq |x|$ for all $|x| \in L/D$. Assume $|x| \leq |y|$, $|y| \leq |x|$. Then $x \to y$, $y \to x \in D$, whence $(x \to y) \odot (y \to x) \in D$. Thus $x \sim_D y$, and therefore $|x| = |y|$. If $|x| \leq |y|$, $|y| \leq |z|$, then $x \to y$, $y \to z \in D$, thus $(x \to y) \odot (y \to z) \leq x \to z \in D$. We conclude $|x| \leq |z|$ and consequently $\leq$ is an order relation on L/D. The elements $|\mathbf{0}|, |\mathbf{1}|$ are the least and the greatest with respect to this order as, for all $x \in L$, $\mathbf{0} \leq x \leq \mathbf{1}$, whence $\mathbf{0} \to x$, $x \to \mathbf{1} \in D$, hence $|\mathbf{0}| \leq |x| \leq |\mathbf{1}|$.□

Theorem 4 *Let D be a ds of a BL-algebra L. Then*

$$\langle L/D, \leq, \wedge, \vee, \odot, \to, |\mathbf{0}|, |\mathbf{1}| \rangle$$

is a BL-algebra, where $|x| \wedge |y| = |x \wedge y|$, $|x| \vee |y| = |x \vee y|$, $|x| \odot |y| = |x \odot y|$, $|x| \to |y| = |x \to y|$.

Proof. First we prove the lattice properties of $\langle L/D, \leq, \wedge, \vee\rangle$. Assume $x, y \in L$. Since $(x \wedge y) \to x = \mathbf{1}$, $(x \wedge y) \to y = \mathbf{1}$ and $\mathbf{1} \in D$, we have $|x \wedge y| \leq |x|, |y|$. Assume now $a \in L$, $|a| \leq |x|, |y|$. Then we have $a \to x \in D, a \to y \in D$, therefore $(a \to x) \wedge (a \to y) \in D$. By (1.64), $(a \to x) \wedge (a \to y) = a \to (x \wedge y)$, thus $|a| \leq |x \wedge y|$ and so $|x \wedge y| =$ g.l.b.$\{|x|, |y|\}$. By a symmetric manner we demonstrate $|x|, |y| \leq |x \vee y|$. If $|x|, |y| \leq |a|$, then $x \to a, y \to a \in D$, so by (1.65), $(x \to a) \wedge (y \to a) = (x \vee y) \to a \in D$, thus $|x \vee y| \leq |a|$ and therefore $|x \vee y| =$ l.u.b.$\{|x|, |y|\}$. The operation $\odot$ on L/D has the property $|x| \odot |\mathbf{1}| = |x \odot \mathbf{1}| = |x|$. It is also associative and commutative as the corresponding operation $\odot$ on L is associative and commutative. To see that $\odot$ on L/D is isotone, let $|x| \leq |y|$. Then $x \to y \in D$. Assume $z \in L$. Since $y \leq z \to (y \odot z)$, we have $x \to y \leq x \to [z \to (y \odot z)] = (x \odot z) \to (y \odot z)$. Therefore $(x \odot z) \to (y \odot z) \in D$, whence $|x| \odot |z| \leq |y| \odot |z|$. The Galois correspondence between $\odot$ and $\to$ on L/D we establish by realizing that $|x| \odot |y| \leq |z|$ iff $(x \odot y) \to z \in D$ iff $x \to (y \to z) \in D$ iff $|x| \leq |y| \to |z|$. Thus, L/D is a residuated lattice. The BL-algebra properties (1.70)-(1.72) follow from those of L.□

Proposition 30 *L/D is linear iff D is a prime ds.*

Proof. If D is a prime *ds* then, by Remark 8, for all $a, b \in L$, either $a \to b \in D$ or $b \to a \in D$. Thus $|a| \leq |b|$ or $|b| \leq |a|$ and therefore L/D is linear. If conversely L/D is linear then, for all $a, b \in L$, either $|a| \leq |b|$ or $|b| \leq |a|$, whence either $a \to b \in D$ or $b \to a \in D$. Assume $a \to b \in D$ and $a \vee b \in D$. Since $a \vee b = [(a \to b) \to b] \wedge [(b \to a) \to a] \leq (a \to b) \to b$, we have $(a \to b) \to b \in D$, thus $b \in D$. Similarly, $b \to a \in D$ implies $a \in D$. Therefore D is prime.□

Definition 8 *The* order *of an element x of a BL-algebra L, in symbols $ord(x)$, is the least integer m such that $x^m = \overbrace{x \odot \ldots \odot x}^{m \text{ terms}} = \mathbf{0}$. If no such m exists then $ord(x) = \infty$. An BL-algebra L is called* locally finite *if all non-unit elements in it are of finite order.*

We immediately realize that Łukasiewicz structure is locally finite while Gödel structure and Product structure are not.

Proposition 31 *For any element x of a BL-algebra L, there is a proper ds D of L such that $x \in D$ iff $ord(x) = \infty$.*

Proof. Let D be a proper *ds* and $x \in D$. Then $x^n \in D$ for any natural number n, whence $x^n \neq \mathbf{0}$ for any $n \in \mathcal{N}$. Therefore $ord(x) = \infty$. Conversely, if $ord(x) = \infty$ then a set $\langle x\rangle = \{y \in L \mid x^n \leq y$ for some $n \in \mathcal{N}\}$ is a *ds* of L. Indeed, trivially $\mathbf{1} \in \langle x\rangle$ and if $y, y \to z \in \langle x\rangle$ then there are natural numbers m, n such that $x^n \leq y$, $x^m \leq (y \to z)$. Since $x^{n+m} \leq (y \to z) \odot y \leq z$, $z \in \langle x\rangle$. Therefore $\langle x\rangle$ is a *ds* of L and is clearly proper as $\mathbf{0} \notin \langle x\rangle$.□

Proposition 32 *Locally finite BL-algebras are linear.*

Proof. Assume $a \vee b = \mathbf{1}$. Then $\mathbf{1} = [(a \to b) \to b] \wedge [(b \to a) \to a] \leq [(a \to b) \to b]$, thus $(b \leq)\ a \to b \leq b$, hence $b = a \to b$. Let now $a \neq \mathbf{1}$. Since the BL-algebra L under consideration is locally finite, there is an m such that $a^m = \mathbf{0}$. Now $b = a \to b = a \to (a \to b) = a^2 \to b = \ldots = a^m \to b = \mathbf{0} \to b = \mathbf{1}$. Thus, $a \vee b = \mathbf{1}$ iff $a = \mathbf{1}$ or $b = \mathbf{1}$. Since, for all elements $a, b \in L$, $(a \to b) \vee (b \to a) = \mathbf{1}$, we have that $a \to b = \mathbf{1}$ or $b \to a = \mathbf{1}$. Therefore $a \leq b$ or $b \leq a$.□

We have seen that in any residuated lattice L holds $\mathbf{0}^* = \mathbf{1}$, $\mathbf{1}^* = \mathbf{0}$ and, for any element $x \in L$, $x \leq x^{**}$, where $x^* = \bigvee\{y \in L \mid x \odot y = \mathbf{0}\}$ by definition. In locally finite BL-algebras, in particular, we have

Proposition 33 *In a locally finite BL-algebra L, for all $x \in L$,*

$$\mathbf{0} < x < \mathbf{1} \quad \textit{iff} \quad \mathbf{0} < x^* < \mathbf{1}, \tag{1.75}$$

$$x^* = \mathbf{0} \quad \textit{iff} \quad x = \mathbf{1}, \tag{1.76}$$

$$x^* = \mathbf{1} \quad \textit{iff} \quad x = \mathbf{0}. \tag{1.77}$$

Proof. Assume $\mathbf{0} < x < \mathbf{1}$, $ord(x) = m\ (\geq 2)$. Then $x^{m-1} \odot x = \mathbf{0}$, $x^{m-2} \odot x \neq \mathbf{0}$ so, by the definition of x^*, $\mathbf{0} < x^{m-1} \leq x^* < x^{m-2} \leq \mathbf{1}$. Conversely, let $\mathbf{0} < x^* < \mathbf{1}$, $ord(x^*) = n(\geq 2)$. Then, by a similar argument, $\mathbf{0} < (x^*)^{n-1} \leq x^{**} < (x^*)^{n-2} \leq \mathbf{1}$. If now $x = \mathbf{0}$, then $x^* = \mathbf{1}$, a contradiction. Therefore $\mathbf{0} < x \leq x^{**} < \mathbf{1}$ and (1.75) is proved. If $x^* = \mathbf{0}$ but $x \neq \mathbf{1}$, then $\mathbf{0} < x < \mathbf{1}$, which leads to a contradiction $x^* \neq \mathbf{0}$. Thus $x = \mathbf{1}$, which proves (1.76). The verification of (1.77) is similar.□

Proposition 34 *In any BL-algebra L, for all $x, y, z \in L$,*

$$(x \to y) \to (x \to z) = (x \wedge y) \to z, \tag{1.78}$$

$$\textit{if } z \to x = z \to y,\ x, y \leq z \textit{ then } x = y, \tag{1.79}$$

$$\textit{if } L \textit{ is linear and } z \to x = z \to y \neq \mathbf{1} \textit{ then } x = y. \tag{1.80}$$

Proof. Let $x, y, z \in L$. Then

$$(x \wedge y) \to z = [(x \to y) \odot x] \to z = (x \to y) \to (x \to z),$$

which proves (1.78). If $x, y \leq z$ then $x = (z \wedge x) = z \odot (z \to x) = z \odot (z \to y) = (z \wedge y) = y$, thus (1.79) holds and, if $z \to x = z \to y \neq \mathbf{1}$, then $z \not\leq x$, $z \not\leq y$ therefore, if L is linear then $x, y \leq z$ and (1.80) now follows by (1.79).□

The following theorem will have a lot of important consequences.

Theorem 5 *$x^{**} = x$ for any element x of a locally finite BL-algebra.*

Proof. It is enough to show that $x^{**} = x$ holds for any element x of a locally finite BL-algebra L and such that $\mathbf{0} < x < \mathbf{1}$; for such an element x we have, by (1.75), $\mathbf{0} < x^* < \mathbf{1}$ and $\mathbf{0} < x^{**} < \mathbf{1}$. By taking $z = \mathbf{0}$ in (1.47) we see, for any $y \in L$, that $(x \odot y)^* = x \to y^*$. Since $x \leq x^{**}$, we have $x = x \wedge x^{**} = x^{**} \odot (x^{**} \to x)$, thus $x^* = [x^{**} \odot (x^{**} \to x)]^* = x^{**} \to (x^{**} \to x)^*$. On the other hand, $x^* = x^{***} = x^{**} \to \mathbf{0}$. Since L is linear and $x^{**} \to \mathbf{0} = x^{**} \to (x^{**} \to x)^* \neq \mathbf{1}$ we have, by (1.80), that $(x^{**} \to x)^* = \mathbf{0}$ and, by (1.76), $x^{**} \to x = \mathbf{1}$. Thus, $x^{**} = x$.□

In Proposition 30 we gave necessary and sufficient condition for L/D to be linear. Now we describe those BL-algebras L/D which are locally finite. We have

Theorem 6 *Let M be a ds of a BL-algebra L. Then the following conditions are equivalent*

$$M \textit{ is a maximal ds.} \tag{1.81}$$

$$\forall x \notin M : \exists n \in \mathcal{N} \textit{ such that } (x^n)^* \in M. \tag{1.82}$$

$$L/M \textit{ is locally finite.} \tag{1.83}$$

Proof. Assume (1.81). Let $x \notin M$. Define a subset D of L by

$$D = \{z \in L \mid \text{for some } y \in M, n \in \mathcal{N}, y \odot x^n \leq z\}.$$

Then trivially $\mathbf{1} \in D$. If $a, a \to b \in D$ then, for some $y, y' \in M$, $n, m \in \mathcal{N}$, holds $y \odot x^n \leq a$, $y' \odot x^m \leq a \to b$. Since $y \odot y' \in M$ and $(y \odot x^n) \odot (y' \odot x^m) = (y \odot y') \odot x^{n+m} \leq a \odot (a \to b) \leq b$, we conclude that $b \in D$ and, therefore, D is a *ds*. Since, for any $y \in M$, $y \odot x \leq y$, we have $M \subseteq D$. But, as $\mathbf{1} \in M$ and $\mathbf{1} \odot x \leq x$, we also have $x \in D$. Since M is maximal this implies $D = L$. Therefore $\mathbf{0} \in D$, i.e. there exists $y \in M$, $n \in \mathcal{N}$ such that $y \odot x^n \leq \mathbf{0}$, in other words $y \leq (x^n)^*$. Hence $(x^n)^* \in M$. Thus, (1.82) holds. Assume now (1.82). Let $|x|_M \in L/M$ be such that $|x|_M \neq |\mathbf{1}|_M$, so $x \notin M$. Then there exists a natural number n such that $(x^n)^* \in M$ and therefore $|x^n|^*_M = |(x^n)^*|_M = |\mathbf{1}|_M$, so that $|x^n|_M \leq |(x^n)^{**}|_M = |\mathbf{1}^*|_M = |\mathbf{0}|_M$. Therefore $|x^n|_M = |x|^n_M = |\mathbf{0}|_M$, whence L/M is locally finite. Finally, assume (1.83). Let D be a *ds* such that $M \subseteq D$. Assume there is an element $x \in L$ such that $x \in D$, $x \notin M$. Then $|x|_M \neq |\mathbf{1}|_M$ and therefore $|x^n|_M = |\mathbf{0}|_M$ for some n, i.e. $\mathbf{0} \sim_M x^n$. Since $M \subseteq D$, also $\mathbf{0} \sim_D x^n$, i.e. $|x^n|_D = |\mathbf{0}|_D$. On the other hand $x \in D$ so $x^n \in D$, thus $|x^n|_D = |\mathbf{1}|_D$, therefore $|\mathbf{0}|_D = |\mathbf{1}|_D$, which implies $\mathbf{0} \in D$, whence M is maximal.□

Theorems 6 and 5 together tell us that, given a maximal *ds* M, the corresponding quotient algebra L/M posses a property $|x|^{**} = |x|$, $|x| \in L/M$.

Proposition 35 *If, in a BL-algebra L, $z^{**} = z$ for all $z \in L$ then, for all $x, y \in L$, $x \vee y = (y \to x) \to x$.*

Proof. We reason that $x \vee y = (x \vee y)^{**} = (x^* \wedge y^*)^* = [x^* \odot (x^* \to y^*)]^* = (x^* \to y^*) \to x^{**} = (y \to x) \to x^{**} = (y \to x) \to x$, where the second equation follows by Exercise 1.3.19, the fourth equation by (1.47) (take $z = \mathbf{0}$), and the fifth equation by Exercise 1.3.18.□

By Proposition 35, if $x^{**} = x$ holds for all $x \in L$ then, for all $x, y \in L$, $(y \to x) \to x = x \vee y = (x \to y) \to y$. BL-algebras of this kind will turn out to be so-called *MV-algebras.* They are in the scope of interest in the next Chapter. Now we study locally finite BL-algebras fulfilling some extra conditions. First we set

Definition 9 *Let L, K be two BL-algebras. A map $h : L \mapsto K$, defined on L, is called a* BL-homomorphism *if, for all $x, y \in L$, $h(x \to y) = h(x) \to h(y)$, $h(x \odot y) = h(x) \odot h(y)$ and $h(\mathbf{0}_L) = \mathbf{0}_K$.*

We leave as an exercise for the reader to prove the following

Remark 10 *If $h : L \mapsto K$ is a BL-homomorphism then, for all $x, y \in L$,*

$$h(x^*) = h(x)^*, h(\mathbf{1}_L) = \mathbf{1}_K, \tag{1.84}$$

$$\text{if } x \leq y \text{ then } h(x) \leq h(y), \tag{1.85}$$

$$h(x \wedge y) = h(x) \wedge h(y), h(x \vee y) = h(x) \vee h(y), \tag{1.86}$$

$$\text{if } D \text{ is ds of } L, \text{ then } h(D) \text{ is ds of } K. \tag{1.87}$$

By (1.86), any BL-homomorphism is also a lattice homomorphism. A BL-homomorphism h is called a *BL-monomorphism* if, whenever $x \neq y$, then $h(x) \neq h(y)$ and h is called a *BL-epimorphism* if, for any $y \in K$, there is $x \in L$ such that $h(x) = y$. An BL-homomorphism which is simultaneously a BL-monomorphism and a BL-epimorphism is called a *BL-isomorphism.* If there is a BL-isomorphism between L and K, then L and K are *isomorphic.*

Given a natural number $m \geq 2$, we introduce an $(m+1)$-element algebraic system $S(m)$, called *finite Łukasiewicz chain,* in the following way

$$S(m) = \{0 = \tfrac{0}{m} < \tfrac{1}{m} < \tfrac{2}{m} < \ldots < \tfrac{m-1}{m} < \tfrac{m}{m} = 1\}.$$

For any $k, p \leq m$, we set

$$\tfrac{p}{m} \odot \tfrac{k}{m} = \tfrac{\max\{0, p+k-m\}}{m}, \; \tfrac{p}{m} \to \tfrac{k}{m} = \tfrac{\min\{0, m-p+k\}}{m}, \; [\tfrac{p}{m}]^* = \tfrac{m-p}{m}.$$

It will be an exercise to demonstrate that any system $S(m)$, $m \geq 2$, is a locally finite BL-algebra.

Proposition 36 *If a BL-algebra L is finitely generated by one element $y \in L$, that is, if $L = \{0 = y^m < y^{m-1} < \ldots < y^1 < y^0 = 1\}$, then L and $S(m)$ are isomorphic.*

Proof. Since the operation $\odot$ is obviously defined, for all $p, k \leq m$, by $y^p \odot y^k = y^{\min\{m,p+k\}}$, it is easy to see that the corresponding residuum is given by $y^p \to y^k = y^{\max\{0,k-p\}}$, in particular $[y^p]^* = y^{m-p}$. Now it is routine to verify that a map $h : L \mapsto S(m)$, given by $h(y^p) = \frac{m-p}{m}$, is a BL-isomorphism.□

Definition 10 *An element $y \neq \mathbf{1}$ of a BL-algebra L is called a* co-atom *if, whenever $y \leq x \leq \mathbf{1}$, then either $x = y$ or $x = \mathbf{1}$.*

The following theorem describes completely those locally finite BL-algebras which contain a co-atom.

Theorem 7 *If a locally finite BL-algebra L contains a co-atom y and $ord(y) = m$, then L and $S(m)$ are isomorphic.*

Proof. Trivially $\mathbf{0} = y^m < y^{m-1} < \ldots < y^1 < y^0 = \mathbf{1}$. We show that any $x \in L$ is of the form y^n for some $n \leq m$. We may assume $\mathbf{0} < x < \mathbf{1}$. Since L is locally finite it is linearly ordered. Thus, for some $n \leq m - 1$ holds $y^{n+1} < x \leq y^n$. The condition $y^{n+1} < x$ implies $y \leq y^n \to x$. Assume $y = y^n \to x$. Since $y^n \not\leq y^{n+1}$, we have $y^n \to y^{n+1} \neq \mathbf{1}$. On the other hand $y \leq y^n \to y^{n+1} \neq \mathbf{1}$, thus $y = y^n \to y^{n+1}$ as y is a co-atom. Therefore $y = y^n \to x = y^n \to y^{n+1}$. It follows from (1.80) that $y^{n+1} = x$, a contradiction. Therefore $y < y^n \to x$ and, as y is a co-atom, $y^n \to x = \mathbf{1}$. Thus $y^n \leq x (\leq y^n)$, whence $y^n = x$. The claim now follows by Proposition 36.□

We will later see that any locally finite BL-algebra is isomorphic to a subalgebra (that is, a subset which is closed with respect to the BL-operations) of the Łukasiewicz structure. By Remark 7, the only proper deductive system of a locally finite BL-algebra is $\{\mathbf{1}\}$. We conclude this section by studying deductive systems of a general BL-algebra L and generated by certain subsets of L. We set

Definition 11 *Given a non-void set A of a BL-algebra L, a set*

$$^{\perp}A = \{x \in L \mid a \vee x = \mathbf{1} \text{ for all } a \in A\}$$

is called a co-annihilator *of A.*

Proposition 37 *$^{\perp}A$ is a ds of L. If $A \neq \{\mathbf{1}\}$ then $^{\perp}A$ is proper.*

Proof. Trivially $\mathbf{1} \in {}^{\perp}A$. Assume $x, x \to y \in {}^{\perp}A$. Let $a \in A$. Then $x \to y \leq x \to (y \vee a)$, $a \leq y \vee a = \mathbf{1} \to (y \vee a) = (x \vee a) \to (y \vee a)$. Therefore

$$\begin{aligned}
\mathbf{1} &= (x \to y) \vee a \\
&\leq [x \to (y \vee a)] \vee [(x \vee a) \to (y \vee a)] \\
&\leq [x \wedge (x \vee a)] \to (y \vee a)] \\
&= x \to (y \vee a),
\end{aligned}$$

where we utilized (1.66) and (1.13). Thus $x \leq y \vee a$, so $\mathbf{1} = x \vee a \leq (y \vee a) \vee a = y \vee a$. Hence $y \in {}^{\perp}A$, whence ${}^{\perp}A$ is a *ds*. If $A \neq \{\mathbf{1}\}$ and as A is non-void, there is an element $x \in A$ such that $x \neq \mathbf{1}$ and $\mathbf{0} \vee x = x \neq \mathbf{1}$. Therefore $\mathbf{0} \notin {}^{\perp}A$ and so ${}^{\perp}A$ is proper.□

Proposition 38 *If A is a linear ds of a BL-algebra L and $A \neq \{\mathbf{1}\}$ then ${}^{\perp}A$ is a prime ds.*

Proof. Assume $a \vee b \in {}^{\perp}A$, but $a \notin {}^{\perp}A$, $b \notin {}^{\perp}A$. Then there exists $x', x'' \in A$ such that $a \vee x' \neq \mathbf{1}$, $b \vee x'' \neq \mathbf{1}$. Set $x = x' \wedge x''$. Then $x \in A$ as A is a *ds* and, clearly, $a \vee x \neq \mathbf{1}$, $b \vee x \neq \mathbf{1}$. Since $x \leq a \vee x$, $b \vee x$, we conclude $a \vee x, b \vee x \in A$ and, as A is linear, we may assume $b \vee x \leq a \vee x$. Now

$$\mathbf{1} = (a \vee b) \vee x = a \vee (b \vee x) \leq a \vee (a \vee x) = a \vee x,$$

which contradicts the fact $a \vee x \neq \mathbf{1}$. Therefore $a \in {}^{\perp}A$ or $b \in {}^{\perp}A$, hence ${}^{\perp}A$ is prime.□

We will see that also the converse of Proposition 38 holds. Given $a \in L$, define

$$D^a = \{x \in L \mid x \to a = a, a \to x = x\}.$$

Then trivially $x \in D^a$ iff $a \in D^x$.

Proposition 39 *For any $a \in L$, D^a is a ds of L as $D^a = {}^{\perp}\{a\}$.*

Proof. If $a \vee x = \mathbf{1}$ then, by (1.71), $(a \to x) \to x = \mathbf{1}$, $(x \to a) \to a = \mathbf{1}$, i.e. $a \to x = x$ and $x \to a = a$. Therefore ${}^{\perp}\{a\} = \{x \in L \mid a \vee x = \mathbf{1}\} = \{x \in L \mid x \to a = a, a \to x = x\} = D^a$.□

Proposition 40 *For any element a of a BL-algebra L,*

$$\text{if } ord(a) < \infty \text{ then } D^a = \{\mathbf{1}\}, \tag{1.88}$$

$$D^a \cap \langle a \rangle = \{\mathbf{1}\}. \tag{1.89}$$

Proof. Assume $ord(a) = m < \infty$, $x \in D^a$. Then $a \in D^x$ which is a *ds*. Therefore $a^m \in D^x$, hence $\mathbf{0} \in D^x$ and so $\mathbf{0} \to x = x$. This means $x = \mathbf{1}$ and so (1.88) is proved. Assume then $x \in D^a \cap \langle a \rangle$. Since $x \in \langle a \rangle$ there is a natural number n such that $a^n \leq x$ and hence, $a^n \to x = \mathbf{1}$. Since $x \in D^a$, we have $a \in D^x$ and therefore also $a^n \in D^x$. Thus $a^n \to x = x$, whence $x = \mathbf{1}$. The proof is complete.□

Remark 11 *For any non-void set $A \subseteq L$, ${}^{\perp}A = \bigcap_{a \in A} D^a = \bigcap_{a \in A} {}^{\perp}\{a\}$.*

Proof. Exercise.

Proposition 41 *If ${}^{\perp}A$ is a prime ds and $a, b \in A$ then either for all $x \in A$ holds $x \in D^{a \to b}$ or for all $x \in A$ holds $x \in D^{b \to a}$*

Proof. Since $(a \to b) \vee (b \to a) = \mathbf{1}$ and ${}^{\perp}A$ is a prime *ds*, either $a \to b \in {}^{\perp}A = \bigcap_{x \in A} D^x$ or $b \to a \in \bigcap_{x \in A} D^x$, thus either for all $x \in A$ holds $x \in D^{a \to b}$ or for all $x \in A$ holds $x \in D^{b \to a}$.□

Now we are able to rewrite Proposition 38.

Proposition 42 *Let $A \subseteq L$ be a ds. Then ${}^{\perp}A$ is a prime ds if, and only if A is linear and $A \neq \{\mathbf{1}\}$.*

Proof. Assume ${}^{\perp}A$ is prime. Then $A \neq \{\mathbf{1}\}$ as otherwise we would have ${}^{\perp}A = L$. Let $a, b \in A$. Since $b \leq a \to b$, $a \leq b \to a$ and A is a *ds*, we have $a \to b, b \to a \in A$. By Proposition 41, either $a \to b, b \to a \in D^{a \to b}$ or $a \to b, b \to a \in D^{b \to a}$. In the first case $\mathbf{1} = (a \to b) \vee (a \to b) = (a \to b)$, thus $a \leq b$, in the second case $\mathbf{1} = (b \to a) \vee (b \to a) = (b \to a)$, hence $b \leq a$. Therefore A is linear. The other direction is Proposition 38.□

Remark 12 *If $X \subseteq Y$, then ${}^{\perp}Y \subseteq {}^{\perp}X$.*

Proof. Exercise.

We generalize the notion $\langle x \rangle$ to concern any non-void set X by defining $\langle X \rangle = \{y \in L \mid x_1 \odot \ldots \odot x_n \leq y \text{ for some } x_1, \ldots, x_n \in X\}$.

Proposition 43 *For a subset $\emptyset \neq X$ of a BL-algebra L, $\langle X \rangle \cap {}^{\perp}X = \{\mathbf{1}\}$.*

Proof. If $a \in \langle X \rangle \cap {}^{\perp}X$, then $a \in {}^{\perp}X$, hence $x \to a = a$ for all $x \in X$. Since $a \in \langle X \rangle$, $x_1 \odot \ldots \odot x_n \leq a$ for some $x_1, \ldots, x_n \in X$, thus

$$\mathbf{1} = x_1 \to (\ldots x_{n-1} \to (x_n \to a) \ldots).$$

But $(x_n \to a) = a$, so $[x_{n-1} \to (x_n \to a)] = a$, ..., finally $\mathbf{1} = a$.□

Proposition 44 *If D is a ds of a BL-algebra L, then $D \cap {}^{\perp}D = \{\mathbf{1}\}$.*

Proof. Let D be a *ds*. We demonstrate that $D = \langle D \rangle$. Trivially $D \subseteq \langle D \rangle$. If $a \in \langle D \rangle$, then $x_1 \odot \ldots \odot x_n \leq a$ for some $x_1, \ldots, x_n \in D$ and $x_1 \odot \ldots \odot x_n \in D$, so $a \in D$ whence $\langle D \rangle = D$. The result now follows by Proposition 43.□

Proposition 45 *If $\emptyset \neq X \subseteq L$, where L is a BL-algebra, then*

$$X \subseteq {}^{\perp\perp}X, \tag{1.90}$$

$${}^{\perp}X = {}^{\perp\perp\perp}X, \tag{1.91}$$

$${}^{\perp}X = {}^{\perp}\langle X \rangle. \tag{1.92}$$

Proof. ${}^{\perp\perp}X = \{a \in L \mid a \vee x = \mathbf{1}$ for all $x \in {}^{\perp}X\}$. If now $b \in X$, then $b \vee x = \mathbf{1}$ for all $x \in {}^{\perp}X$, hence $b \in {}^{\perp\perp}X$ and (1.90) holds. By (1.90), ${}^{\perp}X \subseteq {}^{\perp\perp\perp}X$ and, by Remark 12, ${}^{\perp\perp\perp}X \subseteq {}^{\perp}X$. Therefore (1.91) holds. To verify (1.92), we first reason that a fact $X \subseteq \langle X \rangle$ implies ${}^{\perp}\langle X \rangle \subseteq {}^{\perp}X$. To see that also the converse inclusion holds, assume $y \in {}^{\perp}X$. Then, for any $x_i \in X, i = 1, \ldots n, x_i \vee y = \mathbf{1}$. We demonstrate, by induction on n, that also $(x_1 \odot \ldots \odot x_n) \vee y = \mathbf{1}$. If $n = 1$, then the claim is clearly true, so assume it is true for $n = k$. Let $n = k + 1$ and set $x = (x_1 \odot \ldots \odot x_k)$. By induction hypothesis,

$$\begin{aligned}
\mathbf{1} &= (y \vee x) \odot (y \vee x_{k+1}) \\
&= [y \odot (y \vee x_{k+1})] \vee [x \odot (y \vee x_{k+1})] \\
&= y \vee [(x \odot y) \vee (x \odot x_{k+1})] \\
&\leq y \vee [y \vee (x_1 \odot \ldots \odot x_{k+1})] \\
&= y \vee (x_1 \odot \ldots \odot x_{k+1}).
\end{aligned}$$

Thus, the claim holds for all natural numbers n. If now $z \in \langle X \rangle$ then, for some $x_1, \ldots, x_n \in X$, $x_1 \odot \ldots \odot x_n \leq z$. Therefore $\mathbf{1} = y \vee (x_1 \odot \ldots \odot x_n) \leq y \vee z$. We conclude that $y \vee z = \mathbf{1}$ for any $z \in \langle X \rangle$ and so $y \in {}^{\perp}\langle X \rangle$. This proves ${}^{\perp}X \subseteq {}^{\perp}\langle X \rangle$ and the proof is complete.□

Proposition 46 *If a linear ds D of a BL-algebra L contains an element $x \neq \mathbf{1}$ and $x \vee x^* = \mathbf{1}$, then x is the least element of D.*

Proof. First we realize that $x \vee x^* = \mathbf{1}$ implies $x \wedge x^* = \mathbf{0}$. Indeed, $x \wedge x^* \leq x^{**} \wedge x^* = (x^* \vee x)^* = \mathbf{1}^* = \mathbf{0}$, where the second equation follows by Exercise 1.3.19. Let now $a \in D$. Then $a = a \vee \mathbf{0} = a \vee (x \wedge x^*) = (a \vee x) \wedge (a \vee x^*)$, where the last equation follows by the distributivity of L. By Proposition 42, ${}^{\perp}D$ is a prime *ds*. Since $x \vee x^* = \mathbf{1}$, either $x \in {}^{\perp}D$ or $x^* \in {}^{\perp}D$ and, as $x \vee x = x \neq \mathbf{1}$, we necessarily have $x^* \in {}^{\perp}D$. Now $a \in D$, hence $a \vee x^* = \mathbf{1}$, whence $a = a \vee x$, thus $x \leq a$ and the proof is complete.□

1.5 Exercises

1.1. Lattices and equivalence relations

Exercise 1 Let A by the set of all human beings. Define a binary relation R on A by xRy if person x understands person's y language. Is R a quasi-order on A?

Exercise 2 Prove Remark 1.

Exercise 3 Let A be the set of all independent states in the world. Define a binary relation R on A by
(a) xRy if the number of inhabitants of a state y is at least as big as the number of inhabitants of a state x. Do we obtain a chain?
(b) xRy if the number of inhabitants and the territory of a state y are both at least as big as the number of inhabitants and the territory of a sate x. Does A become a chain under R?

Exercise 4 Show, in details, that the family E of the open sets of the real interval is an partially ordered set under the relation $\subseteq$.

Exercise 5 Prove Remark 2.

Exercise 6 Prove Remark 3.

Exercise 7 Show that in a lattice L, for any $x, y, z \in L$, if $y \leq z$ then $x \vee y \leq x \vee z$.

Exercise 8 Prove that equation (1.16) implies equation (1.15).

Exercise 9 Demonstrate that $\langle E, \subseteq, \cap, \cup \rangle$, where E is the family of the open sets of the real interval, is a distributive lattice.

Exercise 10 Show that any chain is a distributive lattice.

Exercise 11 Associate in the distributive lattice $\langle I, \leq, \min, \max \rangle$ to each element $x \in I$ the element $x^* = 1 - x$. Then $\langle I, \leq, \min, \max, ^* \rangle$ is the structure of the set of values of truth in some fuzzy inference systems. It is also present in 'Fuzzy Sets', the first paper of fuzzy theory published by Zadeh in 1965. Is this structure a Boolean algebra?

Exercise 12 Prove that in a Boolean algebra there is the least element $\mathbf{0}$ and the greatest element $\mathbf{1}$.

Exercise 13 Show that an l-complement x^* of an element x of a Boolean algebra is unique.

Exercise 14 A distributive lattice $\langle L, \leq, \wedge, \vee \rangle$ with a unary operation * such that, for each $x, y \in L$,

$$x^*, y^* \in L \text{ and } (x \wedge y)^* = x^* \vee y^*,\ (x \vee y)^* = x^* \wedge y^*$$

is called a *soft algebra*. Prove that any Boolean algebra is a soft algebra.

Exercise 15 Prove that the distributive lattice $\langle I, \leq, \min, \max, ^* \rangle$ is a soft algebra. (See Exercise 14)

1.2. Lattice filters

Exercise 1 Prove Remark 5.

Exercise 2 Let A, B be two lattices, B containing the unit element $\mathbf{1}_B$. Let F be a filter of B. Assume a mapping $h : A \to B$ is such that, for any $a, b \in A$, $h(a \wedge b) = h(a) \wedge h(b)$, $h(a \vee b) = h(a) \vee h(b)$. Then h is called a *lattice homomorphism.* Prove that
(a) $G = \{a \in A \mid h(a) = \mathbf{1}_B\}$ is a filter of A and (b) $H = \{a \in A \mid h(a) \in F\}$ is a filter of A provided that these sets are non-void.

1.3. Residuated lattices

Exercise 1 Given a generalized residuated lattice $\langle L \leq, \wedge, \vee, \mu, f, g \rangle$, prove that g is antitone in the first and isotone in the second variable.

Exercise 2 Let A be an partially ordered set containing the unit $\mathbf{1}$. Assume binary operations $\odot$ and $\to$, defined on A, satisfy (i) $\odot$ is isotone, (ii) $\to$ is antitone in the first and isotone in the second variable, and (iii) for all $x, y, z \in A$, $x \odot y \leq z$ iff $x \leq y \to z$. Prove that the following are couples of equivalent conditions:

(A)	$(x \odot y) \odot z \leq x \odot (y \odot z)$	(A')	$y \to z \leq (x \to y) \to (x \to z)$
(B)	$x \odot \bigvee_{i \in \Gamma} y_i = \bigvee_{i \in \Gamma} (x \odot y_i)$	(B')	$\bigvee_{i \in \Gamma} y_i \to x = \bigwedge_{i \in \Gamma} (y_i \to x)$
(C)	$x \odot \mathbf{1} = x$	(C')	$x = \mathbf{1} \to x$
(D)	$\mathbf{1} \odot x = x$	(D')	$x \leq y$ iff $\mathbf{1} = x \to y$
(E)	$x \odot y = y \odot x$	(E')	$x \leq y \to z$ iff $y \leq x \to z$
(F)	$x \odot (y \odot z) \leq (x \odot y) \odot z$	(F')	$x \to y \leq (x \odot z) \to (y \odot z)$
(G)	$(x \odot y) \odot z = x \odot (y \odot z)$	(G')	$(x \odot y) \to z = x \to (y \to z)$

Exercise 3 Let L be a complete lattice and $\odot$ an isotone, associative and commutative binary operation defined on L. Assume, moreover, that (1.28) holds. Now define another binary operation $\to$ in L by equation (1.29). Prove, without any extra assumptions, that (a) the Galois correspondence (1.27) holds, (b) the operation $\to$ is antitone in the first variable, and (c) the operation $\to$ is isotone in the second variable.

Exercise 4 Let L be a complete lattice and $\to$ a binary operation on L which is antitone in the first and isotone in the second variable. Moreover, assume that $x \to \bigwedge_{i \in \Gamma} y_i = \bigwedge_{i \in \Gamma} (x \to y_i)$ holds for each $x \in L$, $\{y_i\}_{i \in \Gamma} \subseteq L$. Define another binary operation $\odot$ in L by $x \odot y = \bigwedge \{z \mid x \leq y \to z\}$. Prove that (a) $\odot$ is isotone (b) the Galois correspondence (1.27) holds.

Exercise 5 Consider the structure of Example 1. Let p be a fixed natural number. Prove that the binary operation $\odot$, defined by (1.30), is associative, commutative and isotone and $x \odot \mathbf{1} = x$ holds for any real number $x \in I$. What is the connection between this structure and generalized Łukasiewicz structure?

Exercise 6 Prove that Brouwer structure is a residuated lattice.

Exercise 7 Prove the Gaines structure is a residuated lattice.

Exercise 8 Prove that any Boolean algebra can be regarded as a residuated lattice where the adjoin couple is $\langle \wedge, \to \rangle$, the residuum is defined by $x \to y = x^* \vee y$ and x^* is the l-complement of x. Verify that complement and l-complement coincide in Boolean algebra.

Exercise 9 Prove equations (1.34) - (1.46) and (1.48).

Exercise 10 Show that the order relation $\leq$ in (1.51) and (1.54) cannot, in general, be replaced by $=$.

Exercise 11 Verify equations (1.56), (1.57) and (1.58).

Exercise 12 Show that in a linear residuated lattice L, for all elements $x, y \in L$, $x \leftrightarrow y = (x \vee y) \to (x \wedge y)$.

Exercise 13 Prove that in a residuated lattice L, for all $x, y, z \in L$, $x \leq y \to z$ iff $y \leq x \to z$.

Exercise 14 Prove that in a residuated lattice L holds, for all $x, y, \in L$, $x \odot (y \to z) \leq y \to (x \odot z)$.

Exercise 15 Find an example of a complete residuated lattice L such that

$$\bigvee_{i \in \Gamma} (y_i \to x) < \bigwedge_{i \in \Gamma} y_i \to x$$

or

$$\bigvee_{i \in \Gamma} (x \to y_i) < x \to \bigvee_{i \in \Gamma} y_i$$

hold for some element $x \in L$ and some subset $\{y_i\}_{i \in \Gamma} \subseteq L$.

Exercise 16 A binary operation $t : I \times I \to I$ is called *T-norm*[7] if, for all elements $x, y, z \in I$,
(i) if $x \leq y$, then $t(x, z) \leq t(y, z)$,
(ii) $t(x, y) = t(y, x)$,
(iii) $t(x, 1) = x$,
(iv) $t(x, t(y, z)) = t(t(x, y), x)$,
(v) $t(x, 0) = 0$.
Prove that the product operation $\odot$ of any residuated lattice defined on the unit interval I is a T-norm.

Exercise 17 Verify equations (1.64) and (1.65).

Exercise 18 Assume $x^{**} = x$ holds for all elements x of a residuated lattice L. Prove, for all $x, y \in L$, that $x \to y = y^* \to x^*$.

Exercise 19 Verify $(x \vee y)^* = x^* \wedge y^*$ for all elements x, y of a residuated lattice L.

Exercise 20 Prove $x \odot \bigwedge_{i \in \Gamma} y_i \leq \bigwedge_{i \in \Gamma}(x \odot y_i)$ for all elements x and all subsets $\{y_i\}_{i \in \Gamma}$ of a complete residuated lattice L.

Exercise 21 Verify $(x \to y) \odot z \leq x \to (y \odot z)$ for all elements x, y, z of a residuated lattice L.

1.4. BL-Algebras

Exercise 1 A T-norm t (see Exercise 1.3.16) is called *continuous* if it is a continuous function on the unit interval. Prove that any continuous T-norm t generates a BL-algebra $\langle [0, 1], \leq \min, \max, t, \to_t, 0, 1 \rangle$, where the residuum $\to_t$ is defined by (1.29). In the light of this results, study Łukasiewicz, Gödel and Product structures.

Exercise 2 Define on the unit interval, for all $x, y \in [0, 1]$,

$$x \odot y = \begin{cases} 0 & \text{if } x + y \leq 1 \\ x \wedge y & \text{elsewhere} \end{cases}, \quad x \to y = \begin{cases} 1 & \text{if } x \leq y \\ (1 - x) \wedge y & \text{elsewhere} \end{cases}$$

Do we obtain (a) a residuated lattice (b) a BL-algebra?

Exercise 3 Let $\Rightarrow$ denote the Brouwer residuum. Prove that, for all $x, y \in [0, 1]$, $x \Leftrightarrow y \leq x \leftrightarrow_t y$, where t is a T-norm and $\leftrightarrow_t$ is the corresponding bi-residuum.

[7] T-norms are frequently used in various connections in fuzzy logic framework.

Exercise 4 Prove that D is a prime *ds* of a BL-algebra L iff, for all $a, b \in L$, either $a \to b \in D$ or $b \to a \in D$. Show also that L is linear iff each *ds* D of L is prime.

Exercise 5 Define the proper, prime and maximal deductive systems of (a) Łukasiewicz structure, (b) Brouwer structure and (c) Gaines structure. Compare them with the lattice filters of the unit interval.

Exercise 6 Prove Remark 9.

Exercise 7 Show that $D \neq \emptyset$ is a *ds* of an BL-algebra L iff (i) if $a, b \in D$ then $a \odot b \in D$ and (ii) if $a \in D, a \leq b$ then $b \in D$.

Exercise 8 Prove that a set $D = \{a \in L \mid a \sim \mathbf{1}\}$, where $\sim$ is a congruence relation on a BL-algebra L, is a *ds* of L.

Exercise 9 Prove that $x \sim_D y$ iff $(x \to y) \odot (y \to x) \in D$, where D is a *ds* of a BL-algebra L, is a congruence relation with respect to $\wedge$ and $\vee$.

Exercise 10 Complete the proof of Proposition 28 by showing that there is a one-to-one correspondence between the maximal deductive systems of a BL-algebra L and the maximal congruence relations on L.

Exercise 11 Prove Remark 10.

Exercise 12 Verify the fact that any algebraic system $S(m), m \geq 1$, is a locally finite BL-algebra.

Exercise 13 Prove Remark 11.

Exercise 14 Prove Remark 12.

Chapter 2

MV-Algebras

2.1 MV-Algebras and Wajsberg Algebras

Mathematicians want to minimize the set of axioms of a certain mathematical theory and maximize the set of consequences of these axioms. In the previous chapter we show that in this respect, in the theory of residuated lattices, the Galois correspondence is an extremely effective axiom. In this chapter we introduce Wajsberg[1] algebras, which will have a lot of important consequences each having direct application in fuzzy logic. In fact, Wajsberg algebras are in the core of fuzzy set theory. We also study MV-algebras[2] by giving first a rather long definition of this algebraic structure. Our definition for MV-algebra will not be the most economical but shows some basic important properties this structure obeys. We also prove that there is one-to-one correspondence between MV-algebras and Wajsberg algebras; each MV-algebra can be seen as a Wajsberg algebra and vice versa. MV-algebras will turn out to be particular BL-algebras, thus residuated lattices, a fact that justifies our notation.

Definition 12 *Let L be a non-void set, $\mathbf{1} \in L$ and $\rightarrow$, * be a binary and a unary operation, respectively, defined on L such that, for each $x, y, z \in L$,*

$$\mathbf{1} \rightarrow x = x, \tag{2.1}$$

$$(x \rightarrow y) \rightarrow [(y \rightarrow z) \rightarrow (x \rightarrow z)] = \mathbf{1}, \tag{2.2}$$

$$(x \rightarrow y) \rightarrow y = (y \rightarrow x) \rightarrow x, \tag{2.3}$$

$$(x^* \rightarrow y^*) \rightarrow (y \rightarrow x) = \mathbf{1}. \tag{2.4}$$

Then the system $\langle L, \rightarrow, ^*, \mathbf{1}\rangle$ is called a Wajsberg algebra.

[1] Wajsberg was the first (1935) to show that the infinite-valued Łukasiewicz propositional calculus is complete with respect to the axioms conjectured by Łukasiewicz. Unfortunately, Wajsberg's proof was never published.

[2] Chang introduced MV-algebras in 1958. MV-algebras allowed him to give another completeness proof for Łukasiewicz logic.

Define on a Wajsberg algebra $\langle L, \rightarrow, ^*, \mathbf{1}\rangle$ a binary relation $\leq$ by

$$x \leq y \text{ iff } x \rightarrow y = \mathbf{1}. \tag{2.5}$$

Then $\leq$ is an order relation on L and $\mathbf{1}$ is the greatest element in L, in other words, for each $x, y, z \in L$,

$$x \rightarrow x = \mathbf{1}, \tag{2.6}$$
$$\text{if } x \rightarrow y = y \rightarrow z = \mathbf{1}, \text{ then } x \rightarrow z = \mathbf{1}, \tag{2.7}$$
$$\text{if } x \rightarrow y = y \rightarrow x = \mathbf{1}, \text{ then } x = y, \tag{2.8}$$
$$x \rightarrow \mathbf{1} = \mathbf{1}. \tag{2.9}$$

Indeed, by (2.2), $(\mathbf{1} \rightarrow \mathbf{1}) \rightarrow [(\mathbf{1} \rightarrow x) \rightarrow (\mathbf{1} \rightarrow x)] = \mathbf{1}$, hence by (2.1), $\mathbf{1} \rightarrow (x \rightarrow x) = \mathbf{1}$, thus by (2.1), $x \rightarrow x = \mathbf{1}$. Therefore $x \leq x$. Let $x \leq y$, $y \leq z$, i.e. $x \rightarrow y = y \rightarrow z = \mathbf{1}$. By (2.2),

$$(x \rightarrow y) \rightarrow [(y \rightarrow z) \rightarrow (x \rightarrow z)] = \mathbf{1},$$

thus $\mathbf{1} \rightarrow [\mathbf{1} \rightarrow (x \rightarrow z)] = \mathbf{1}$, which implies $x \rightarrow z = \mathbf{1}$. Therefore $x \leq z$. Let $x \leq y$, $y \leq x$, i.e. $x \rightarrow y = y \rightarrow x = \mathbf{1}$. By (2.3), $(x \rightarrow y) \rightarrow y = (y \rightarrow x) \rightarrow x$, thus, $\mathbf{1} \rightarrow y = \mathbf{1} \rightarrow x$ and therefore $x = y$. We conclude that (2.5) defines an order on L. To see that $\mathbf{1}$ is the greatest element in L with respect to this order, we first reason, by (2.3), (2.1) and (2.1), that $(x \rightarrow \mathbf{1}) \rightarrow \mathbf{1} = (\mathbf{1} \rightarrow x) \rightarrow x = x \rightarrow x = \mathbf{1}$. Therefore

$$(x \rightarrow \mathbf{1}) \rightarrow \mathbf{1} = \mathbf{1}. \tag{2.10}$$

On the other hand, by (2.1), (2.10), (2.1) and (2.2), $\mathbf{1} \rightarrow (x \rightarrow \mathbf{1}) = x \rightarrow \mathbf{1}$ $= x \rightarrow [(x \rightarrow \mathbf{1}) \rightarrow \mathbf{1}] = (\mathbf{1} \rightarrow x) \rightarrow [(x \rightarrow \mathbf{1}) \rightarrow (\mathbf{1} \rightarrow \mathbf{1})] = \mathbf{1}$, in other words, $\mathbf{1} \rightarrow (x \rightarrow \mathbf{1}) = \mathbf{1}$. Thus, $(x \rightarrow \mathbf{1}) \rightarrow \mathbf{1} = \mathbf{1} \rightarrow (x \rightarrow \mathbf{1}) = \mathbf{1}$ so, by (2.8), $x \rightarrow \mathbf{1} = \mathbf{1}$, i.e. $x \leq \mathbf{1}$ for any element $x \in L$.

Remark 13 *In a Wajsberg algebra $\langle L, \rightarrow, ^*, \mathbf{1}\rangle$, the operation $\rightarrow$ is antitone in the first variable.*

Proof. Exercise.

Proposition 47 *In a Wajsberg algebra $\langle L, \rightarrow, ^*, \mathbf{1}\rangle$, for any $x, y, z \in L$,*

$$x \rightarrow (y \rightarrow x) = \mathbf{1}, \tag{2.11}$$
$$(x \rightarrow y) \rightarrow [(z \rightarrow x) \rightarrow (z \rightarrow y)] = \mathbf{1}, \tag{2.12}$$
$$x \rightarrow (y \rightarrow z) = y \rightarrow (x \rightarrow z). \tag{2.13}$$

Proof. Let $x, y, z \in L$. Since $y \leq \mathbf{1}$ and $\rightarrow$ is antitone in the first variable we reason that $\mathbf{1} \rightarrow x \leq y \rightarrow x$, therefore $x \leq y \rightarrow x$, thus $x \rightarrow (y \rightarrow x) = \mathbf{1}$. Hence (2.11) holds. To establish (2.12), we first verify

$$\text{if } x \rightarrow (y \rightarrow z) = \mathbf{1}, \text{ then } y \rightarrow (x \rightarrow z) = \mathbf{1}. \tag{2.14}$$

Indeed, if $x \leq y \rightarrow z$ then, as the operation $\rightarrow$ is antitone in the first variable, $(y \rightarrow z) \rightarrow z \leq x \rightarrow z$. By (2.3), $(z \rightarrow y) \rightarrow y \leq x \rightarrow z$ and, by (2.11), $y \leq (z \rightarrow y) \rightarrow y$, so that altogether we have $y \leq x \rightarrow z$. Therefore (2.14) holds. Applying now (2.14) to $(z \rightarrow x) \rightarrow [(x \rightarrow y) \rightarrow (z \rightarrow y)] = \mathbf{1}$, which holds by (2.2), we conclude that $(x \rightarrow y) \rightarrow [(z \rightarrow x) \rightarrow (z \rightarrow y)] = \mathbf{1}$, i.e. (2.12) holds. Finally, by (2.11) and (2.3),

$$y \leq (z \rightarrow y) \rightarrow y = (y \rightarrow z) \rightarrow z,$$

and by (2.12), $(y \rightarrow z) \rightarrow z \leq [x \rightarrow (y \rightarrow z)] \rightarrow (x \rightarrow z)$. Therefore $y \leq [x \rightarrow (y \rightarrow z)] \rightarrow (x \rightarrow z)$ which, by (2.14), implies $x \rightarrow (y \rightarrow z) \leq y \rightarrow (x \rightarrow z)$. By a symmetric argument $y \rightarrow (x \rightarrow z) \leq x \rightarrow (y \rightarrow z)$. We conclude that (2.13) holds. □

Remark 14 *In a Wajsberg algebra $\langle L, \rightarrow, ^*, \mathbf{1} \rangle$, the operation $\rightarrow$ is isotone in the second variable.*

Proof. Exercise.

Proposition 48 *In a Wajsberg algebra $\langle L, \rightarrow, ^*, \mathbf{1} \rangle$, for any $x \in L$,*

$$x^* = x \rightarrow \mathbf{1}^*, \tag{2.15}$$

$$\mathbf{1}^* \leq x. \tag{2.16}$$

Proof. By (2.11), $x^* \leq (\mathbf{1}^*)^* \rightarrow x^*$ and, by (2.4), $(\mathbf{1}^*)^* \rightarrow x^* \leq x \rightarrow \mathbf{1}^*$, therefore

$$x^* \leq x \rightarrow \mathbf{1}^*. \tag{2.17}$$

On the other hand, by (2.4) and (2.1), $x^* \rightarrow \mathbf{1}^* \leq \mathbf{1} \rightarrow x = x$, and as $\rightarrow$ is antitone in the first variable, $x \rightarrow \mathbf{1}^* \leq (x^* \rightarrow \mathbf{1}^*) \rightarrow \mathbf{1}^*$ which, by (2.3), implies that $x \rightarrow \mathbf{1}^* \leq (\mathbf{1}^* \rightarrow x^*) \rightarrow x^*$ and, by (2.13),

$$\mathbf{1}^* \rightarrow x^* \leq (x \rightarrow \mathbf{1}^*) \rightarrow x^*. \tag{2.18}$$

Moreover, we have $\mathbf{1}^* \leq x^*$. Indeed, by (2.4) and (2.1), $(x^*)^* \rightarrow \mathbf{1}^* \leq \mathbf{1} \rightarrow x^* = x^*$ and, by (2.11), $\mathbf{1}^* \leq (x^*)^* \rightarrow \mathbf{1}^*$, thus $\mathbf{1}^* \leq x^*$. Hence, $\mathbf{1}^* \rightarrow x^* = \mathbf{1}$ and, by (2.18), $\mathbf{1} = (x \rightarrow \mathbf{1}^*) \rightarrow x^*$. Therefore

$$x \rightarrow \mathbf{1}^* \leq x^*. \tag{2.19}$$

(2.17) and (2.19) now imply (2.15). By (2.11), (2.4) and (2.1) we have $\mathbf{1}^* \leq x^* \rightarrow \mathbf{1}^* \leq \mathbf{1} \rightarrow x = x$. Therefore (2.16) holds. □

For simplicity, we will sometimes denote a Wajsberg algebra $\langle L, \rightarrow, ^*, \mathbf{1} \rangle$ by L.

Remark 15 *Condition* (2.16) *implies that* $\mathbf{1}^*$ *is the least element in a Wajsberg algebra* L *and will therefore be denoted by* $\mathbf{0}$. *As was the case with residuated lattices, we will write* x^{**} *instead of* $(x^*)^*$. *We leave an exercise for the reader to verify that for any elements* $x, y, z \in L$,

$$x^{**} = x, \tag{2.20}$$

$$x^* \to y^* = y \to x, \tag{2.21}$$

$$x \leq y \text{ iff } y^* \leq x^*. \tag{2.22}$$

Define two binary operations $\wedge$ and $\vee$ on a Wajsberg algebra L by

$$x \vee y = (x \to y) \to y, \tag{2.23}$$

$$x \wedge y = (x^* \vee y^*)^*. \tag{2.24}$$

Proposition 49 *For any Wajsberg algebra* L, *the system* $\langle L, \leq, \wedge, \vee \rangle$ *is a lattice, where the operations* $\leq$, $\vee$ *and* $\wedge$ *are defined by* (2.5), (2.23) *and* (2.24), *respectively.*

Proof. We have to show that l.u.b$\{x, y\} = (x \to y) \to y$. By (2.11), $y \leq (x \to y) \to y$ and, by (2.11), (2.3), $x \leq (y \to x) \to x = (x \to y) \to y$. Let now $z \in L$ be such that x, $y \leq z$. Then $x \to z = \mathbf{1}$ so, by (2.1), $(x \to z) \to z = z$. Since $\to$ is antitone in the first variable, we reason first $z \to x \leq y \to x$ and then $(y \to x) \to x \leq (z \to x) \to x$. By (2.3), $(y \to x) \to x \leq (x \to z) \to z = z$, thus $(x \to y) \to y \leq z$. We conclude that $x \vee y$, defined by (2.23), coincide with l.u.b$\{x, y\}$. To prove that g.l.b$\{x, y\}$ $= (x^* \vee y^*)^*$, we first realize that $(x^* \vee y^*)^* \leq x, y$ iff $x^*, y^* \leq x^* \vee y^*$, which is the case. On the other hand, if $z \in L$ is such that $z \leq x$, y, then x^*, y^* $\leq z^*$, thus $x^* \vee y^* \leq z^*$ and so $z \leq (x^* \vee y^*)^*$. We conclude that (2.24) is a correct definition. □

Remark 16 *In the lattice* $\langle L, \leq, \wedge, \vee \rangle$ *as defined above, we have, for each* $x, y \in L$,

$$(x \wedge y)^* = x^* \vee y^*, (x \vee y)^* = x^* \wedge y^*. \tag{2.25}$$

Proof. Exercise.

Define on a Wajsberg algebra L a binary operation $\odot$, for each $x, y \in L$, via

$$x \odot y = (x \to y^*)^*. \tag{2.26}$$

Then we have

Proposition 50 *For each $x, y, z \in L$,*

$$x \odot y = y \odot x, \tag{2.27}$$

$$x \odot (y \odot z) = (x \odot y) \odot z, \tag{2.28}$$

$$\text{if } x \leq y, \text{ then } x \odot z \leq y \odot z, \tag{2.29}$$

$$x \odot \mathbf{1} = x, \tag{2.30}$$

$$x \odot y \leq z \text{ iff } x \leq y \to z. \tag{2.31}$$

Proof. By (2.26), (2.20), (2.21), (2.26), respectively, $x \odot y = (x \to y^*)^* = (x^{**} \to y^*)^* = (y \to x^*)^* = y \odot x$, hence (2.27) holds. Next we reason

$$\begin{array}{rcll} x \odot (y \odot z) & = & x \odot (z \odot y) & \text{(by (2.27))} \\ & = & x \odot (z \to y^*)^* & \text{(by (2.26))} \\ & = & [x \to (z \to y^*)^{**}]^* & \text{(by (2.26))} \\ & = & [x \to (z \to y^*)]^* & \text{(by (2.20))} \\ & = & [z \to (x \to y^*)]^* & \text{(by (2.13))} \\ & = & [z \to (x \to y^*)^{**}]^* & \text{(by (2.20))} \\ & = & z \odot (x \to y^*)^* & \text{(by (2.26))} \\ & = & z \odot (x \odot y) & \text{(by (2.26))} \\ & = & (x \odot y) \odot z. & \text{(by (2.27))} \end{array}$$

Thus (2.28) holds. If $x \leq y$ then, as $\to$ is isotone in the first variable, $y \to z^* \leq x \to z^*$, so $(x \to z^*)^* \leq (y \to z^*)^*$. Therefore $x \odot z \leq y \odot z$. We conclude that (2.29) holds. (2.30) and (2.31) are left as exercises for the reader. □

We now summarize our results thus far on Wajsberg algebras by writing

Theorem 8 *Given a Wajsberg algebra $\langle L, \to, ^*, \mathbf{1} \rangle$, the algebraic system $\langle L, \leq, \wedge, \vee, \odot, \to, \mathbf{0}, \mathbf{1} \rangle$, where the operations $\leq, \wedge, \vee, \odot$ are defined by* (2.5), (2.24), (2.23) *and* (2.26), *respectively, and $\mathbf{0} = \mathbf{1}^*$, is a residuated lattice.*

Thus, all properties valid in any residuated lattice hold in Wajsberg algebras, too. For example, by (1.65) and (1.64), for each $x, y, z \in L$

$$(x \vee y) \to z = (x \to z) \wedge (y \to z), \tag{2.32}$$

$$x \to (y \wedge z) = (x \to y) \wedge (x \to z). \tag{2.33}$$

By Theorem 8, any Wajsberg algebra can viewed as a residuated lattice. In general, the converse is not true. In Exercise 1.3.10 we studied such a residuated lattice that (2.3) and (2.20), true in every Wajsberg algebra, do not hold.

We will later give necessary and sufficient conditions for a residuated lattice to be a Wajsberg algebra. Now we study some further lattice properties of Wajsberg algebras.

Proposition 51 *In a Wajsberg algebra L, for each $x, y, z \in L$,*

$$(x \to y) \vee (y \to x) = \mathbf{1}, \tag{2.34}$$
$$(x \wedge y) \to z = (x \to y) \to (x \to z). \tag{2.35}$$

Proof. By (2.32), $(x \vee y) \to x = (x \to x) \wedge (y \to x) = y \to x$ and, similarly $(x \vee y) \to y = x \to y$. Therefore

$$\begin{aligned}
&(y \to x) \to (x \to y) \\
&\quad = [(x \vee y) \to x] \to [(x \vee y) \to y] \\
&\quad = [x^* \to (x \vee y)^*] \to [y^* \to (x \vee y)^* && \text{(by (2.21))} \\
&\quad = y^* \to \{[x^* \to (x \vee y)^*] \to (x \vee y)^*\} && \text{(by (2.13))} \\
&\quad = y^* \to [x^* \vee (x \vee y)^*] && \text{(by (2.23))} \\
&\quad = [x^* \vee (x \vee y)^*]^* \to y && \text{(by (2.21),(2.20))} \\
&\quad = [x \wedge (y \vee x)] \to y && \text{(by (2.25),(2.20))} \\
&\quad = x \to y. && \text{(by (2.14))}
\end{aligned}$$

Thus $\mathbf{1} = [(y \to x) \to (x \to y)] \to (x \to y) = (y \to x) \vee (x \to y)$. We conclude that (2.34) holds. (2.35) will be an exercise for the reader. The proof is complete. □

Define on a Wajsberg algebra L a binary operation $\oplus$, for $x, y \in L$, by

$$x \oplus y = x^* \to y, \tag{2.36}$$

Then we have

$$x \wedge y = (x \oplus y^*) \odot y, \tag{2.37}$$
$$x \vee y = (x \odot y^*) \oplus y. \tag{2.38}$$

Indeed, we realize

$$\begin{aligned}
x \wedge y &= (x^* \vee y^*)^* && \text{(by (2.24))} \\
&= [(x^* \to y^*) \to y^*]^* && \text{(by (2.23))} \\
&= [y \to (x^* \to y^*)^*]^* && \text{(by (2.21))} \\
&= y \odot (x^* \to y^*) && \text{(by (2.26))} \\
&= y \odot (x \oplus y^*) && \text{(by (2.36))} \\
&= (x \oplus y^*) \odot y && \text{(by (2.27))}
\end{aligned}$$

and, moreover,

$$\begin{aligned}
(x \odot y^*) \oplus y &= (x \odot y^*)^* \to y && \text{(by (2.36))} \\
&= (x \to y^{**})^{**} \to y && \text{(by (2.26))} \\
&= (x \to y) \to y && \text{(by (2.20))} \\
&= x \vee y. && \text{(by (2.23))}
\end{aligned}$$

We leave for the reader to verify that the following equations under the given notation are satisfied in every Wajsberg algebra.

$$x \oplus y = y \oplus x \quad , \quad x \odot y = y \odot x, \tag{2.39}$$
$$x \oplus (y \oplus z) = (x \oplus y) \oplus z \quad , \quad x \odot (y \odot z) = (x \odot y) \odot z, \tag{2.40}$$
$$x \oplus x^* = \mathbf{1} \quad , \quad x \odot x^* = \mathbf{0}, \tag{2.41}$$
$$x \oplus \mathbf{1} = \mathbf{1} \quad , \quad x \odot \mathbf{0} = \mathbf{0}, \tag{2.42}$$
$$x \oplus \mathbf{0} = x \quad , \quad x \odot \mathbf{1} = x, \tag{2.43}$$
$$(x \oplus y)^* = x^* \odot y^* \quad , \quad (x \odot y)^* = x^* \oplus y^*, \tag{2.44}$$
$$x^{**} = x \quad , \quad \mathbf{1}^* = \mathbf{0}, \tag{2.45}$$
$$x \vee y = y \vee x \quad , \quad x \wedge y = y \wedge x, \tag{2.46}$$
$$x \vee (y \vee z) = (x \vee y) \vee z \quad , \quad x \wedge (y \wedge z) = (x \wedge y) \wedge z, \tag{2.47}$$
$$x \oplus (y \wedge z) = (x \oplus y) \wedge (x \oplus z) \quad , \quad x \odot (y \vee z) = (x \odot y) \vee (x \odot z) \tag{2.48}$$

Definition 13 *An algebraic system* $\langle L, \oplus, \odot, ^*, \mathbf{0}, \mathbf{1} \rangle$ *which satisfies* (2.39) - (2.48) *and where the operations* $\wedge$ *and* $\vee$ *are defined by* (2.37) *and* (2.38), *respectively, is called an* MV-algebra.

We immediately realize that any Wajsberg algebra $\langle L, \rightarrow, ^*, \mathbf{1} \rangle$ defines an MV-algebra structure. Also the converse is true. Indeed, given an MV-algebra $\langle L, \oplus, \odot, ^*, \mathbf{0}, \mathbf{1} \rangle$, we can define an operation $\rightarrow$ such that the Wajsberg algebra axioms hold. To see this, we first establish

Proposition 52 *An MV-algebra* $\langle L, \oplus, \odot, ^*, \mathbf{0}, \mathbf{1} \rangle$ *satisfies*

$$\mathbf{0}^* = \mathbf{1}, \tag{2.49}$$
$$\langle L, \leq, \wedge, \vee \rangle \text{ is a lattice}, \tag{2.50}$$

where the order relation $\leq$ *is defined by*

$$x \leq y \text{ iff } x \vee y = x \text{ iff } x \wedge y = y \text{ iff } x^* \oplus y = \mathbf{1} \text{ iff } x \odot y^* = \mathbf{0}.$$

Moreover, $\mathbf{0}$ *is the least element and* $\mathbf{1}$ *is the greatest element with respect to this order relation and, for each* $x, y, z \in L$,

$$(x \vee y)^* = x^* \wedge y^* \text{ and } (x \wedge y)^* = x^* \vee y^*, \tag{2.51}$$
$$\text{if } x \leq y, \text{ then } x \oplus z \leq y \oplus z, x \odot z \leq y \odot z, \tag{2.52}$$
$$x \odot y \leq x \wedge y \leq x, y \leq x \vee y \leq x \oplus y. \tag{2.53}$$

Proof. By (2.45), $\mathbf{0}^* = (\mathbf{1}^*)^* = \mathbf{1}$. To establish (2.50) we need some preliminary results. Without referring to particular MV-axioms we reason, for each $x, y, z \in L$, that $x \vee \mathbf{0} = (x \odot \mathbf{0}^*) \oplus \mathbf{0} = (x \odot \mathbf{1}) \oplus \mathbf{0} = x \oplus \mathbf{0} = x$, $x \vee \mathbf{1} = (x \odot \mathbf{1}^*) \oplus \mathbf{1} = \mathbf{1}$, $x \wedge \mathbf{1} = (x \oplus \mathbf{1}^*) \odot \mathbf{1} = (x \oplus \mathbf{0}) \odot \mathbf{1} = x \odot \mathbf{1} = x$, $x \wedge \mathbf{0} = (x \oplus \mathbf{0}^*) \odot \mathbf{0} = \mathbf{0}$. Therefore

$$x \vee \mathbf{0} = x = x \wedge \mathbf{1}, x \vee \mathbf{1} = \mathbf{1}, x \wedge \mathbf{0} = \mathbf{0}. \wedge \mathbf{1}. \tag{2.54}$$

Moreover, $x \vee x = (x \odot x^*) \oplus x = \mathbf{0} \oplus x = x$, $x \wedge x = (x \oplus x^*) \odot x = \mathbf{1} \odot x = x$, thus

$$x \vee x = x = x \wedge x. \tag{2.55}$$

Now, $(x \vee y)^* = [(x \odot y^*) \oplus y]^* = (x \odot y^*)^* \odot y^* = (x^* \oplus y^{**}) \odot y^* = x^* \wedge y^*$ and, by a similar argument, $(x \wedge y)^* = x^* \vee y^*$, thus,

$$(x \vee y)^* = x^* \wedge y^*, (x \wedge y)^* = x^* \vee y^*. \tag{2.56}$$

We have established (2.51). Then we reason $x \wedge (x \vee y) = (y \vee x) \wedge x = [(y \odot x^*) \oplus x] \wedge x = \{[(y \odot x^*) \oplus x] \oplus x^*\} \odot x = [(y \odot x^*) \oplus (x \oplus x^*)] \odot x = [(y \odot x^*) \oplus \mathbf{1}] \odot x = \mathbf{1} \odot x = x$. Similarly, $x \vee (x \wedge y) = x$. Hence,

$$x \wedge (x \vee y) = x = x \vee (x \wedge y). \tag{2.57}$$

If $x \oplus y = \mathbf{0}$, then $x \wedge \mathbf{0} = x \wedge (x \oplus y) = (x \oplus \mathbf{0}) \wedge (x \oplus y) = x \oplus (\mathbf{0} \wedge y) = x \oplus \mathbf{0} = x$. By a similar argument $y = \mathbf{0}$, too. Thus,

$$\text{if } x \oplus y = \mathbf{0}, \text{ then } y = x = \mathbf{0}. \tag{2.58}$$

By a symmetric argument we verify

$$\text{if } x \odot y = \mathbf{1}, \text{ then } y = x = \mathbf{1}. \tag{2.59}$$

If $x \vee y = \mathbf{0}$, then $(x \odot y^*) \oplus y = \mathbf{0}$, hence $y = \mathbf{0}$ and then $x \odot \mathbf{1} = x = \mathbf{0}$. We conclude that

$$\text{if } x \vee y = \mathbf{0}, \text{ then } y = x = \mathbf{0}. \tag{2.60}$$

Again, by a symmetric argument we establish

$$\text{if } x \wedge y = \mathbf{1}, \text{ then } y = x = \mathbf{1}. \tag{2.61}$$

Now we define a binary relation $\leq$ on L by

$$x \leq y \text{ iff } x \vee y = y. \tag{2.62}$$

Then, by (2.55),

$$x \leq x. \tag{2.63}$$

If $x \leq y$, $y \leq z$, then $x \vee y = y$, $y \vee z = z$ and $x \vee z = x \vee (y \vee z) = (x \vee y) \vee z = y \vee z = z$. Thus $x \leq z$. We conclude

$$\text{if } x \leq y, y \leq z, \text{ then } x \leq z. \tag{2.64}$$

We also realize that

$$\text{if } x \leq y, y \leq x, \text{ then } y = x \vee y = y \vee x = x. \tag{2.65}$$

By (2.54), $\mathbf{0} \vee x = x$, $x \vee \mathbf{1} = \mathbf{1}$, thus

$$\mathbf{0} \leq x \leq \mathbf{1}. \tag{2.66}$$

By (2.63)-(2.66), we conclude that (2.62) defines an order relation on L and $\mathbf{0}$, $\mathbf{1}$ are the least element and the greatest element, respectively, with respect to this order. If $x \leq y$, then $x \vee y = y$ and, by (2.57), $x = x \wedge (x \vee y)$ $= x \wedge y$. Conversely, if $x = x \wedge y$, then $x \vee y = (x \wedge y) \vee y = y$, thus $x \leq y$ and therefore

$$x \leq y \text{ iff } x = x \wedge y, \tag{2.67}$$

$$x \leq y \text{ iff } x = x \wedge y \text{ iff } x^* = (x \wedge y)^* = x^* \vee y^* \text{ iff } y^* \leq x^*. \tag{2.68}$$

If $x \leq y$, then $x = x \wedge y$, $y = x \vee y$ and therefore $(x \vee z) \vee (y \vee z) =$ $(x \vee y) \vee (z \vee z) = y \vee z$, hence $x \vee z \leq y \vee z$. Similarly $x \wedge z \leq y \wedge z$. We write

$$\text{if } x \leq y, \text{ then } x \vee z \leq y \vee z, x \wedge z \leq y \wedge z. \tag{2.69}$$

In particular, $x \wedge y \leq x \wedge \mathbf{1} = x = x \vee \mathbf{0} \leq x \vee y$, thus

$$x \wedge y \leq x, y \leq x \vee y. \tag{2.70}$$

If $z \leq x, y$, then $z = x \wedge z = y \wedge z$ and $z = z \wedge z = (x \wedge z) \wedge (y \wedge z) =$ $(x \wedge y) \wedge z$, therefore $z \leq x \wedge y$. We have

$$\text{if } z \leq x, y, \text{ then } z \leq x \wedge y. \tag{2.71}$$

Similarly,

$$\text{if } x, y \leq z, \text{ then } x \vee y \leq z. \tag{2.72}$$

(2.70)-(2.72) how imply that $\langle L, \leq, \wedge, \vee \rangle$ is a lattice. If $x \leq y$, then $x \wedge y =$ x and $x \vee y = y$. Therefore $(x \oplus z) \wedge (y \oplus z) = (x \wedge y) \oplus z = x \oplus z$. Hence $x \oplus z \leq y \oplus z$ and, by a similar argument, $x \odot z \leq y \odot z$. We write

$$\text{if } x \leq y, \text{ then } x \oplus z \leq y \oplus z, x \odot z \leq y \odot z, \tag{2.73}$$

and conclude that (2.52) holds. In particular, $x \odot y \leq x \odot \mathbf{1} = x = x \oplus \mathbf{0}$ $\leq x \oplus y$, thus

$$x \odot y \leq x, y \leq x \oplus y, \tag{2.74}$$

and, by (2.70),

$$x \odot y \leq x \wedge y \leq x, y \leq x \vee y \leq x \oplus y. \tag{2.75}$$

Therefore (2.53) holds. It now remains to complete the proof of (2.50). Clearly $y \oplus x^* = \mathbf{1}$ iff $x \odot y^* = \mathbf{0}$. If $x \odot y^* = \mathbf{0}$, then $x \vee y = (x \odot y^*) \oplus y = \mathbf{0} \oplus y = y$, hence $x \leq y$. Conversely, if $x \leq y$, then $\mathbf{1} = x \oplus x^* \leq y \oplus x^* \leq \mathbf{1}$, thus $\mathbf{1} = y \oplus x^*$. We conclude

$$x \leq y \text{ iff } x = x \wedge y \text{ iff } y = x \vee y \text{ iff } y \oplus x^* = \mathbf{1} \text{ iff } x \odot y^* = \mathbf{0}. \quad (2.76)$$

Thus, (2.50) holds. The proof in complete. □

If the order relation defined by (2.62) is a total order, then the corresponding MV-algebra is called an *MV-chain*. Now we write

Proposition 53 *Define on an MV-algebra* $\langle L, \oplus, \odot, ^*, \mathbf{0}, \mathbf{1}\rangle$ *a binary operation* $\rightarrow$, *for each* $x, y \in L$, *by*

$$x \rightarrow y = x^* \oplus y. \quad (2.77)$$

Then we obtain a Wajsberg algebra $\langle L, \rightarrow, ^*, \mathbf{1}\rangle$.

Proof. Let $x, y, z \in L$. We show that $\langle L, \rightarrow, ^*, \mathbf{1}\rangle$ satisfies the equations (2.1)-(2.4). Indeed,

$$\begin{aligned} \mathbf{1} \rightarrow x &= \mathbf{1}^* \oplus x && \text{(by (2.77))} \\ &= \mathbf{0} \oplus x && \text{(by (2.45))} \\ &= x \oplus \mathbf{0} && \text{(by (2.39))} \\ &= x. && \text{(by (2.43))} \end{aligned}$$

Therefore (2.1) holds. Moreover,
$(x \rightarrow y) \rightarrow [(y \rightarrow z) \rightarrow (x \rightarrow z)]$

$$\begin{aligned} &= (x^* \oplus y)^* \oplus [(y^* \oplus z)^* \oplus (x^* \oplus z)] && \text{(by (2.77))} \\ &= (x \odot y^*) \oplus [(y \odot z^*) \oplus (x^* \oplus z)] && \text{(by (2.45),(2.44))} \\ &= [(y^* \odot x) \oplus x^*] \oplus [(y \odot z^*) \oplus z] && \text{(by (2.40),(2.39))} \\ &= (y^* \vee x^*) \oplus (y \vee z) && \text{(by (2.38),(2.45))} \\ &\geq y^* \oplus y && \text{(by (2.53))} \\ &= \mathbf{1}. && \text{(by (2.41))} \end{aligned}$$

We conclude that (2.2) holds. Then we make the following observations

$$\begin{aligned} (x \rightarrow y) \rightarrow y &= (x^* \oplus y)^* \oplus y && \text{(by (2.77))} \\ &= (x \odot y^*) \oplus y && \text{(by (2.45), (2.44))} \\ &= x \vee y && \text{(by (2.38))} \\ &= y \vee x && \text{(by (2.46))} \\ &= (y \odot x^*) \oplus x && \text{(by (2.38))} \\ &= (y^* \oplus x)^* \oplus x && \text{(by (2.45), (2.44))} \\ &= (y \rightarrow x) \rightarrow x, && \text{(by (2.77)} \end{aligned}$$

so (2.3) holds. Finally,

$$\begin{aligned} y \to x &= y^* \oplus x && \text{(by (2.77)} \\ &= x \oplus y^* && \text{(by (2.39))} \\ &= x^{**} \oplus y^* && \text{(by (2.45))} \\ &= x^* \to y^*. && \text{(by (2.77))} \end{aligned}$$

Thus, by (2.50), $(x^* \to y^*)^* \oplus (y \to x) = \mathbf{1}$, i.e. $(x^* \to y^*) \to (y \to x) = \mathbf{1}$. Therefore (2.4) holds. The proof is complete. □

We have proved that, given an MV-algebra $\langle L, \odot, \oplus, ^*, \mathbf{0}, \mathbf{1} \rangle$, we obtain a Wajsberg algebra $\langle L, \to, ^*, \mathbf{1} \rangle$ by defining the operation $\to$ via (2.77) and, conversely, a Wajsberg algebra $\langle L, \to, ^*, \mathbf{1} \rangle$ generates an MV-algebra $\langle L, \odot, \oplus, ^*, \mathbf{0}, \mathbf{1} \rangle$, where the operations $\odot$, $\oplus$ are defined by (2.26), (2.36), respectively, and $\mathbf{0} = \mathbf{1}^*$. Thus, we summarize our result by writing

Theorem 9 *There is a one-to-one correspondence between MV-algebras and Wajsberg algebras.*

By Theorem 9, an MV-algebra has all the Wajsberg algebra properties and vice versa. In MV-algebras we have, for example, the following interesting cancellation law:

$$\text{If } x \oplus z = y \oplus z, x \leq z^*, y \leq z^*, \text{ then } x = y. \tag{2.78}$$

Indeed, by assumption, $\mathbf{1} = x^* \oplus z^* = y^* \oplus z^*$. Therefore

$$\begin{aligned} x &= x \odot \mathbf{1} && \text{(by (2.43))} \\ &= x \odot (x^* \oplus z^*) && \text{(by assumption)} \\ &= (z^* \oplus x^*) \odot x && \text{(by (2.39))} \\ &= z^* \wedge x && \text{(by (2.37))} \\ &= x \wedge z^* && \text{(by (2.46))} \\ &= (x \oplus z) \odot z^* && \text{(by (2.37),(2.45))} \\ &= (y \oplus z) \odot z^* && \text{(by (assumption)} \\ &= y \wedge z^* && \text{(by (2.37),(2.45))} \\ &= z^* \wedge y && \text{(by (2.46))} \\ &= (z^* \oplus y^*) \odot y && \text{(by (2.37))} \\ &= \mathbf{1} \odot y && \text{(by assumption)} \\ &= y. && \text{(by (2.43))} \end{aligned}$$

Proposition 54 *MV-algebras are BL-algebras.*

Proof. MV-algebras are residuated lattices where the BL-algebra axioms (1.70) - (1.72) hold by (2.37), (2.23) and (2.34), respectively.□

A direct consequence of Proposition 54 and Remark 16 is

Proposition 55 *The complemented lattice $\langle L, \leq, \wedge, \vee, ^* \rangle$ generated by an MV-algebra L is a soft algebra.*

Now we return to the mutual relation of residuated lattices and MV-algebras. The following theorem gives us a simple criteria to distinguish between MV-algebras and other residuated lattices.

Theorem 10 *A residuated lattice* $\langle L, \leq, \wedge, \vee, \odot, \rightarrow, \mathbf{0}, \mathbf{1}\rangle$ *is a Wajsberg algebra* $\langle L, \rightarrow, ^*, \mathbf{1}\rangle$ *if, and only if it satisfies an additional condition*

$$(x \rightarrow y) \rightarrow y = (y \rightarrow x) \rightarrow x, \text{ for any } x, y \in L, \tag{2.79}$$

with an abbreviation $x^* = x \rightarrow \mathbf{0}$.

Proof. Condition (2.79) is clearly necessary. We show that it is also sufficient. We have to show that, in such a residuated lattice that (2.79) holds, also (2.1)-(2.4) are satisfied. Now, (2.1), (2.2) and (2.3) hold by (1.34), (1.45) and assumption, respectively. Moreover,

$$\begin{aligned} (x \rightarrow \mathbf{0}) \rightarrow \mathbf{0} &= (\mathbf{0} \rightarrow x) \rightarrow x && \text{(by (2.79))} \\ &= \mathbf{1} \rightarrow x && \text{(by (1.43))} \\ &= x. && \text{(by (1.34))} \end{aligned}$$

Thus, $x^{**} = x$ for any $x \in L$. Therefore

$$\begin{aligned} x^* \rightarrow y^* &= (x \rightarrow \mathbf{0}) \rightarrow (y \rightarrow \mathbf{0}) \\ &= y \rightarrow [(x \rightarrow \mathbf{0}) \rightarrow \mathbf{0}] && \text{(by (1.48))} \\ &= y \rightarrow x^{**} \\ &= y \rightarrow x. \end{aligned}$$

We conclude that $(x^* \rightarrow y^*) \rightarrow (y \rightarrow x) = \mathbf{1}$. Thus, (2.4) holds. The proof is complete. □

Notice that in a residuated lattice L such that (2.79) holds, l.u.b.$\{x, y\}$ $= (x \rightarrow y) \rightarrow y$. Indeed, $x \rightarrow y \leq x \rightarrow y$, whence $x \odot (x \rightarrow y) \leq y$, hence $x \leq (x \rightarrow y) \rightarrow y$ and similarly $y \leq (y \rightarrow x) \rightarrow x = (x \rightarrow y) \rightarrow y$. If $x, y \leq z$ then $x \rightarrow z = \mathbf{1}$, so $z = \mathbf{1} \rightarrow z = (x \rightarrow z) \rightarrow z = (z \rightarrow x) \rightarrow x$. Moreover, $z \rightarrow x \leq y \rightarrow x$, therefore $(y \rightarrow x) \rightarrow x \leq (z \rightarrow x) \rightarrow x = z$. Similarly $(x \rightarrow y) \rightarrow y \leq (z \rightarrow x) \rightarrow x$ and we have proved l.u.b.$\{x, y\}$ $= (x \rightarrow y) \rightarrow y$. It will be an exercise to show that if (2.79) holds in a residuated lattice L, then g.l.b$\{x, y\} = x \odot (x \rightarrow y) = y \odot (y \rightarrow x)$. An immediate consequence of Theorem 10 and Proposition 35 is

Theorem 11 *A BL-algebra* L *defines an MV-algebra if, and only if* $x^{**} = x$ *holds for all* $x \in L$.

By Theorem 5 and Theorem 11, we see that any locally finite BL-algebra is an MV-algebra, in particular, the quotient algebra L/D generated by a maximal deductive system D of a BL-algebra L is an MV-algebra.

Theorem 10 states that Wajsberg algebas or equivalently, MV-algebras, are exactly those residuated lattices where $x \vee y$ and $(x \rightarrow y) \rightarrow y$ coincide.

We have seen that Boolean algebras can be regarded as residuated lattices where $\odot$ coincides with $\wedge$, x^* coincides with the l-complement of an element $x \in L$ and $x \to y = x^* \vee y$. We leave for the reader to verify that (2.79) then holds and write

Proposition 56 *Boolean algebras are Wajsberg algebras.*

Proposition 57 *In a Wajsberg algebra L, the following are equivalent conditions*

$$x \odot x = x, \tag{2.80}$$

$$\odot \textit{ and } \wedge \textit{ coincide}, \tag{2.81}$$

$$x \oplus x = x, \tag{2.82}$$

$$\oplus \textit{ and } \vee \textit{ coincide}, \tag{2.83}$$

$$x \vee x^* = \mathbf{1}, \tag{2.84}$$

$$x \wedge x^* = \mathbf{0}, \tag{2.85}$$

$$x \wedge (x \to y) = x \wedge y, \tag{2.86}$$

where x, y are any elements of L.

Proof. (2.81) trivially implies (2.80). Let (2.80) hold. By assumption and since the operations $\wedge$, $\odot$ are isotone and by (2.53), we have $x \wedge y = (x \wedge y) \odot (x \wedge y) \leq x \odot y \leq x \wedge y$. Therefore (2.81) holds. In an symmetric manner we prove that (2.82) and (2.83) are equivalent. The equivalence between (2.80) and (2.82) is left as an exercise for the reader. Since $\mathbf{1} = x \vee x^*$ iff $\mathbf{0} = (x \vee x^*)^* = x^* \wedge x = x \wedge x^*$, (2.84) and (2.85) are equivalent. Assume (2.84) holds. Then, by the MV-algebra axioms and (1.28), $x = x \odot \mathbf{1} = x \odot (x \vee x^*) = (x \odot x) \vee (x \odot x^*) = (x \odot x) \vee \mathbf{0} = x \odot x$. Thus, (2.84) implies (2.80). Let (2.83) hold. Then $x \vee x^* = x \oplus x^* = \mathbf{1}$ by (2.41). Thus, (2.83) implies (2.84). Assume (2.86) holds. Then $x \wedge x^* = x \wedge (x \to \mathbf{0}) = x \wedge \mathbf{0} = \mathbf{0}$. Therefore (2.85) is valid. Finally, assume (2.81) holds. Since $y \leq x \to y$, we have $x \wedge y \leq x \wedge (x \to y)$. On the other hand, $x \wedge (x \to y) = x \odot (x \to y) \leq y$ and $x \wedge (x \to y) \leq x$, therefore $x \wedge (x \to y) \leq x \wedge y$. We conclude that (2.81) implies (2.86). The proof is complete. □

An element $x \in L$ such that any of the conditions (2.80) - (2.86) holds is called *Boolean.* Clearly $\mathbf{0}$, $\mathbf{1}$ are Boolean elements. The set of all Boolean elements of a given Wajsberg algebra L is denoted by BoL. Obviously, BoL is a Boolean algebra. Proposition 56 and Proposition 57 clarify the difference between Wajsberg algebras and Boolean algebras and explain the significance of the operations $\odot$ and $\oplus$. Later, when we will talk about fuzzy logic, we will see that these operations offer a natural interpretation for the logical connectives *conjunction* and *disjunction*, respectively, just in the same way as $\wedge$ and $\vee$ do in classical Boolean logic. We summarize our results by writing

Theorem 12 *A Wajsberg algebra is a Boolean algebra if, and only if it satisfies any of the equations* (2.80)-(2.86).

2.2 Complete MV-Algebras

Recall an MV-algebra is called *complete* if it contains the greatest lower bound and the lowest upper bound of any of its subset. In particular, an MV-algebra L is called *countable complete* if $\bigwedge\{y_i \mid i \in \Gamma\} \in L$ and $\bigvee\{y_i \mid i \in \Gamma\} \in L$ hold for any countable subset $\{y_i \mid i \in \Gamma\}$ of L. It will turn out that all countable complete MV-algebras L are representable by ordinary fuzzy sets, that is, there is a collection of $[0,1]$-valued fuzzy sets structurally isomorphic to L. This fact is of special importance as such MV-algebras are the only BL-algebras possessing such a representation property. By Theorem 5 and Theorem 11 we know that locally finite BL-algebras are MV-algebras and Theorem 7 tells us a locally finite BL-algebra containing an co-atom of order m is isomorphic to $S(m)$ which, in turn, is a subalgebra of Łukasiewicz structure. In this section we will prove that any locally finite BL-algebra is isomorphic to a subalgebra of Łukasiewicz structure. We use quite frequently the basic properties of MV-algebras we established in the previous section and will not cite them every time they are used. To avoid extra parentheses, we let $\odot$ be more binding than $\oplus$. Thus, e.g. $a \oplus b \odot (c \oplus d)$ stands for $a \oplus [b \odot (c \oplus d)]$. We start by establishing some properties of complete MV-algebras.

Proposition 58 *In a complete MV-algebra L, for each $x \in L$, $\{y_i\}_{i\in\Gamma} \subseteq L$,*

$$\bigwedge_{i\in\Gamma}(x \oplus y_i) = x \oplus \bigwedge_{i\in\Gamma} y_i, \tag{2.87}$$

$$\bigwedge_{i\in\Gamma} y_i^* = (\bigvee_{i\in\Gamma} y_i)^*, \tag{2.88}$$

$$\bigvee_{i\in\Gamma} y_i^* = (\bigwedge_{i\in\Gamma} y_i)^*, \tag{2.89}$$

$$\bigwedge_{i\in\Gamma}(x \vee y_i) = x \vee \bigwedge_{i\in\Gamma} y_i, \tag{2.90}$$

$$\bigvee_{i\in\Gamma}(x \oplus y_i) = x \oplus \bigvee_{i\in\Gamma} y_i. \tag{2.91}$$

Proof. Recall that in a Wajsberg algebra L holds, for each $x, y, z \in L$, the Galois correspondence $x \odot y \leq z$ iff $x \leq y^* \oplus z$. Now, for each $i \in \Gamma$, $\bigwedge_{i\in\Gamma}(x \oplus y_i) \leq x \oplus y_i$ iff $[\bigwedge_{i\in\Gamma}(x \oplus y_i)] \odot x^* \leq y_i$ iff $[\bigwedge_{i\in\Gamma}(x \oplus y_i)] \odot x^* \leq \bigwedge_{i\in\Gamma} y_i$ iff $\bigwedge_{i\in\Gamma}(x \oplus y_i) \leq x \oplus \bigwedge_{i\in\Gamma} y_i$. The converse $x \oplus \bigwedge_{i\in\Gamma} y_i. \leq \bigwedge_{i\in\Gamma}(x \oplus y_i)$ holds as the operation $\oplus$ is isotone. We therefore have (2.87). Since $\bigwedge_{i\in\Gamma} y_i^* \leq y_i^*$ for each $i \in \Gamma$, we have $y_i \leq (\bigwedge_{i\in\Gamma} y_i^*)^*$ for each $i \in \Gamma$.

Therefore $\bigvee_{i\in\Gamma} y_i \leq (\bigwedge_{i\in\Gamma} y_i^*)^*$, or equivalently $\bigwedge_{i\in\Gamma} y_i^* \leq (\bigvee_{i\in\Gamma} y_i)^*$. Conversely, since $y_i \leq \bigvee_{i\in\Gamma} y_i$ for each $i \in \Gamma$, we have $(\bigvee_{i\in\Gamma} y_i)^* \leq y_i^*$ for each $i \in \Gamma$, which implies $(\bigvee_{i\in\Gamma} y_i)^* \leq \bigwedge_{i\in\Gamma} y_i^*$. Thus, (2.88) holds. (2.89) is left as an exercise for the reader the reader. Then we reason

$$\begin{aligned}
[x \vee (\textstyle\bigwedge_{i\in\Gamma} y_i)]^* &= x^* \wedge (\textstyle\bigwedge_{i\in\Gamma} y_i)^* && \text{(by (2.51))}\\
&= x^* \wedge (\textstyle\bigvee_{i\in\Gamma} y_i^*) && \text{(by (2.89))}\\
&= \textstyle\bigvee_{i\in\Gamma}(x^* \wedge y_i^*) && \text{(Exercise 2)}\\
&= \textstyle\bigvee_{i\in\Gamma}(x \vee y_i)^*. && \text{(by (2.51))}
\end{aligned}$$

Hence, we have

$$\begin{aligned}
x \vee (\textstyle\bigwedge_{i\in\Gamma} y_i) &= [\textstyle\bigvee_{i\in\Gamma}(x \vee y_i)^*]^* && \text{(by (2.45))}\\
&= \textstyle\bigwedge_{i\in\Gamma}(x \vee y_i)^{**} && \text{(by (2.88))}\\
&= \textstyle\bigwedge_{i\in\Gamma}(x \vee y_i). && \text{(by (2.45))}
\end{aligned}$$

Thus, (2.90) holds. Next we observe

$$\begin{aligned}
[\textstyle\bigwedge_{i\in\Gamma}(y_i \odot x)] \oplus x^* &= \textstyle\bigwedge_{i\in\Gamma}[(y_i \odot x) \oplus x^*] && \text{(by (2.87))}\\
&= \textstyle\bigwedge_{i\in\Gamma}(y_i \vee x^*) && \text{(by (2.38))}\\
&= (\textstyle\bigwedge_{i\in\Gamma} y_i) \vee x^* && \text{(by (2.90))}\\
&= [(\textstyle\bigwedge_{i\in\Gamma} y_i) \odot x] \oplus x^*. && \text{(by (2.38))}
\end{aligned}$$

Since $\bigwedge_{i\in\Gamma}(y_i \odot x) \leq x$, $(\bigwedge_{i\in\Gamma} y_i) \odot x \leq x$ we conclude, by (2.78), that

$$x \odot \bigwedge_{i\in\Gamma} y_i = \bigwedge_{i\in\Gamma}(x \odot y_i) \tag{2.92}$$

holds. Finally,

$$\begin{aligned}
\textstyle\bigvee_{i\in\Gamma}(x \oplus y_i) &= [\textstyle\bigvee_{i\in\Gamma}(x \oplus y_i)]^{**} && \text{(by (2.45))}\\
&= [\textstyle\bigwedge_{i\in\Gamma}(x \oplus y_i)^*]^* && \text{(by (2.88))}\\
&= [\textstyle\bigwedge_{i\in\Gamma}(x^* \odot y_i^*)]^* && \text{(by (2.44))}\\
&= [x^* \odot (\textstyle\bigwedge_{i\in\Gamma} y_i^*)]^* && \text{(by (2.92))}\\
&= [x^* \odot (\textstyle\bigvee_{i\in\Gamma} y_i)^*]^* && \text{(by (2.88))}\\
&= x^{**} \oplus (\textstyle\bigvee_{i\in\Gamma} y_i)^{**} && \text{(by (2.44))}\\
&= x \oplus (\textstyle\bigvee_{i\in\Gamma} y_i). && \text{(by (2.45))}
\end{aligned}$$

We have established (2.91). The proof is complete. □

Definition 14 *A complete residuated lattice L is called* continuous *if, for any $x \in L$, $\{y_i\}_{i\in\Gamma} \subseteq L$,*

$$x \to \bigvee_{i\in\Gamma} y_i = \bigvee_{i\in\Gamma}(x \to y_i), \tag{2.93}$$

$$\bigwedge_{i\in\Gamma} y_i \to x = \bigvee_{i\in\Gamma}(y_i \to x). \tag{2.94}$$

We leave as an exercise for the reader to verify that complete residuated lattices generated by Wajsberg algebras are continuous. The following theorem clarifies the origin of this concept. It also gives us another criteria to distinguish between MV-algebras and other residuated lattices of the unit interval.

Theorem 13 *A complete residuated lattice L defined on the unit interval is continuous*
(i) *if, and only if, for all $a \in I$, the functions f_a, g_a defined, for any $x \in I$, by $f_a(x) = x \to a$, $g_a(x) = a \to x$ are continuos functions,*
(ii) *if, and only if L forms a complete Wajsberg algebra.*

Proof. To prove (i), let $a \in [0,1]$. We show that f_a is continuous at any point $b \in [0,1]$. Assume $b \neq 0$. Since f_a is decreasing, we have, by (1.65), $\lim_{x\to b^-} f_a(x) = \bigwedge_{x<b}(x \to a) = \bigvee_{x<b} x \to a = b \to a = f_a(b)$. Assume $b \neq 1$. By (2.94), $\lim_{x\to b^+} f_a(x) = \bigvee_{b<x}(x \to a) = \bigwedge_{b<x} x \to a = b \to a$ $= f_a(b)$. Thus $\lim_{x\to b} f_a(x) = f_a(b)$ for any $b \neq 0, 1$ and $\lim_{x\to 1^-} f_a(x)$ $= f_a(1)$, $\lim_{x\to 0^+} f_a(x) = f_a(0)$. We conclude that f_a is continuous in the unit interval. The verification of continuity of the functions g_a is left for the reader. Assume the residuated lattice on the unit interval under consideration is complete but some f_a is not a continuous function. Then f_a is not continuous at some point $b \in [0,1]$. Since $\lim_{x\to b^-} f_a(x) = f_a(b)$, we have $b \neq 1$ and $\lim_{x\to b^+} f_a(x) \neq f_a(b)$. Thus, by (1.66), $\lim_{x\to b^+} f_a(x)$ $= \bigvee_{b<x}(x \to a) < \bigwedge_{b<x} x \to a = b \to a = f_a(b)$. Therefore (2.94) does not hold. It is an exercise to verify that discontinuity of some g_a implies that (2.93) does not hold. Therefore the residuated lattice under consideration is not continuous.

To verify (ii), we first assume L is continuous. Then, for any $a \in I$, f_a is a continuous function. We claim

$$(x \to a) \to a = \max\{a, x\}.$$

Indeed, if $x \leq a$ then, by (1.40), (1.34), $(x \to a) \to a = 1 \to a = a =$ $\max\{a, x\}$. Let $a < x \leq 1$. Since $f_a(1) = a$, $f_a(a) = 1$ and f_a is continuous, there is $y \in [a, 1]$ such that $f_a(y) = y \to a = x$. Then

$$(x \to a) \to a = [(y \to a) \to a] \to a \leq y \to a = x = \max\{a, x\},$$

where the middle (in-)equality is obtained by applying the antitonicity of $\to$ in the first variable and the fact $y \leq (y \to a) \to a$. Thus, by Theorem 10, L is a Wajsberg algebra. The converse is left as an exercise for the reader. □

Proposition 59 *In an MV-algebra L, for each $x, y, z \in L$, $(x \oplus z) \odot y \leq$ $x \oplus (y \odot z)$.*

Proof. The claim is equivalent to $(x^* \to z) \odot y \leq x^* \to (y \odot z)$, that is, $(x^* \to z) \odot x^* \odot y \leq z \odot y$, which holds true by the Galois correspondence and the isotonicity of $\odot$. □

Proposition 60 *In an MV-algebra L, for each $x, y, z \in L$, if $x \wedge y^* = \mathbf{0}$, then $(x \oplus z) \odot y = x \oplus (y \odot z)$.*

Proof. Letting $w = [x \oplus (y \odot z)]^* \oplus [(x \oplus z) \odot y]$ we see that, by Proposition 59, it is sufficient to prove that $w = \mathbf{1}$. Now,

$$\begin{aligned} x \oplus w &= [x \oplus x^* \odot (y^* \oplus z^*)] \oplus [(x \oplus z) \odot y] \\ &= [y^* \oplus z^* \oplus (x \odot y \odot x)] \oplus [y \odot (x \oplus z)] \\ &= [z^* \oplus (x \odot y \odot z)] \oplus [y^* \oplus y \odot (x \oplus z)] \\ &= [z^* \oplus (x \odot y \odot z)] \oplus [x \oplus z \oplus (y^* \odot x^* \odot z^*] \\ &= \mathbf{1} \end{aligned}$$

and

$$\begin{aligned} y^* \oplus w &= [y^* \oplus y \odot (x \oplus z)] \oplus [x^* \odot (y^* \oplus z^*)] \\ &= [(x \oplus z) \oplus (y^* \odot x^* \odot z^*)] \oplus [x^* \odot (y^* \oplus z^*)] \\ &= [z \oplus (y^* \odot x^* \odot z^*)] \oplus [x \oplus x^* \odot (y^* \oplus z^*)] \\ &= [z \oplus (y^* \odot x^* \odot z^*)] \oplus [(y^* \oplus z^*) \oplus (x \odot y \odot z)] \\ &= \mathbf{1}. \end{aligned}$$

Therefore $(x \oplus w) \wedge (y^* \oplus w) = \mathbf{1}$. By (2.48) and the hypothesis, $\mathbf{1} = (x \wedge y^*) \oplus w = w$ and the proof is complete. □

Proposition 61 *In an MV-algebra L, for each $x, y, z \in L$, if $x \vee y = \mathbf{1}$, then $(x \odot x) \vee (y \odot y) = \mathbf{1}$.*

Proof. By definition,

$$(x \odot x) \vee (y \odot y) = (x \odot x) \oplus (x^* \oplus x^*) \odot y \odot y.$$

By using the hypothesis $x^* \wedge y^* = \mathbf{0}$ and Proposition 60 twice,

$$(x^* \oplus x^*) \odot y \odot y = (x^* \oplus x^* \odot y) \odot y = x^* \oplus x^* \odot y \odot y.$$

Therefore,

$$\begin{aligned} (x \odot x) \vee (y \odot y) &= (x \odot x \oplus x^*) \oplus x^* \odot y \odot y \\ &= (x^* \odot x^* \oplus x) \oplus x^* \odot y \odot y \\ &= x^* \odot x^* \oplus (x \oplus x^* \odot y \odot y) \\ &= x^* \odot x^* \oplus [y \odot y \oplus x \odot (y^* \oplus y^*)]. \end{aligned}$$

Again, by hypothesis and Proposition 60, $x \odot (y^* \oplus y^*) = x \odot y^* \oplus y^*$. This leads to

$$\begin{aligned} (x \odot x) \vee (y \odot y) &= x^* \odot x^* \oplus y \odot y \oplus x \odot y^* \oplus y^* \\ &= x^* \odot x^* \oplus x \odot y^* \oplus (y \odot y \oplus y^*) \\ &= x^* \odot x^* \oplus x \odot y^* \oplus (y^* \odot y^* \oplus y) \\ &= x^* \odot x^* \oplus y^* \odot y^* \oplus (x \odot y^* \oplus y) \\ &= x^* \odot x^* \oplus y^* \odot y^* \oplus (x \vee y) \\ &= \mathbf{1}, \end{aligned}$$

which is the desired conclusion. □

It will be a more exacting exercise to prove Proposition 61 without using Proposition 60. Recall from the theory of BL-algebras that, for each element x and each natural number m, we abbreviate $x^0 = \mathbf{1}$ and $x^{m+1} = x^m \odot x$. Dealing with MV-algebras we also abbreviate $0x = \mathbf{0}$ and $(m+1)x = mx \oplus x$. We leave for the reader to verify that, for each element x of an MV-algebra L and each non-negative integer n, $(nx)^* = (x^*)^n$, $nx^* = (x^n)^*$, $m(nx) = (mn)x$, $x^{(m+n)} = x^m \odot x^n$, $x^{mn} = (x^m)^n$.

Definition 15 *The* order *of an element x of an MV-algebra L is the least integer n such that $nx = \mathbf{1}$, in symbols $O(x) = n$, and if no such integer n exists, then $O(x) = \infty$. An MV-algebra is called* locally finite *if any non-zero element is of finite order.*

Notice that, in general, $ord(x) \neq O(x)$. In the Łukasiewicz structure, for example, $ord(0.6) = 3$ while $O(0.6) = 2$. We have, however,

Proposition 62 *Locally finite BL-algebras and locally finite MV-algebras coincide.*

Proof. If L is a locally finite BL-algebra then, by Theorem 5 and Theorem 11, L is an MV-algebra. Assume there is a non-zero element x such that $mx < \mathbf{1}$ for any $m \in \mathcal{N}$. Then $x \neq \mathbf{1}$ and $\mathbf{0} < (mx)^* = (x^*)^m$ for any $m \in \mathcal{N}$. But then $\mathbf{0} < x^* < \mathbf{1}$ and $ord(x^*) = \infty$, a contradiction. Therefore L is a locally finite MV-algebra. Conversely, by Proposition 54, a locally finite MV-algebra L is a BL-algebra. Assume there is a non-unit element x such that $\mathbf{0} < x^m$ for all $m \in \mathcal{N}$. Then $x \neq \mathbf{0}$ and $mx^* = (x^m)^* < \mathbf{1}$ for all $m \in \mathcal{N}$ so that $\mathbf{0} < x^* < \mathbf{1}$ and $O(x^*) = \infty$, a contradiction. Therefore L is a locally finite BL-algebra.□

Proposition 63 *Let L be an MV-algebra. For each $x, y \in L$,* (i) *if $x \vee y = \mathbf{1}$ then $x^n \vee y^n = \mathbf{1}$ for any $n \geq 1$, and* (ii) *if $x \wedge y = \mathbf{0}$ then $nx \wedge ny = \mathbf{0}$ for any $n \geq 1$.*

Proof. By an easy induction from Proposition 61, we see that the hypothesis (i) leads to $x^{2^m} \vee y^{2^m} = \mathbf{1}$ for any $m \geq 1$. Since for each $n \geq 1$ there exists an m such that $n \leq 2^m$ and since $x \odot x \leq x \odot \mathbf{1} = x$, we see that $x^{2^m} \leq x^n$ and $y^{2^m} \leq y^n$. Therefore $\mathbf{1} = x^{2^m} \vee y^{2^m} \leq x^n \vee y^n \leq \mathbf{1}$. The verification of (ii) is left for the reader.□

Proposition 64 *In an MV-algebra L, for each $x, y \in L$,* (i) *if $O(x \odot y) < \infty$, then $x \oplus y = \mathbf{1}$, and* (ii) *if $ord(x \oplus y) < \infty$ then $x \odot y = \mathbf{0}$.*

Proof. By the hypothesis (i), for some $n \geq 1$, $n(x \odot y) = \mathbf{1}$. Hence $(x^* \oplus y^*)^n = \mathbf{0}$. By (2.34), (2.36), $(x \oplus y) \vee (x^* \oplus y^*) = \mathbf{1}$, and by Proposition 63,

$$(x \oplus y)^n \vee (x^* \oplus y^*)^n = \mathbf{1}.$$

Thus, $(x \oplus y)^n = \mathbf{1}$. Since $(x \oplus y)^n \leq x \oplus y$ we conclude $x \oplus y = \mathbf{1}$. The verification of (ii) is an exercise.□

A direct consequence of Proposition 64 is the following

Proposition 65 *If L is an MV-algebra, $x \in L$ and $O(x) > 2$ then $O(x^2) = \infty$. If $x \in L$ and $ord(x) > 2$ then $O(2x) = \infty$. If, in particular, $ord(2x^*) < \infty$ then $x \oplus x = \mathbf{1}$.*

By Proposition 32 and Proposition 62 we have

Proposition 66 *Locally finite MV-algebras are linear.*

Proposition 34 can be specified for MV-algebras in the following way.

Proposition 67 *Let L be an MV-chain. For each $x, y, z \in L$, if $x \oplus z = y \oplus z \neq \mathbf{1}$, then $x = y$.*

Proof. An assumption $x \oplus z = y \oplus z = \neq \mathbf{1}$ is equivalent to $z^* \to x = z^* \to y = \neq \mathbf{1}$; the results $x = y$ therefore follows from (1.80).□

Proposition 68 *Let L be an MV-chain. For each $x, y, z \in L$, if $x^* \leq y$, $y \oplus z \neq \mathbf{1}$, then $x \odot (y \oplus z) = (x \odot y) \oplus z$.*

Proof. $[(x \odot y) \oplus z] \oplus x^* = (x \odot y) \oplus x^* \oplus z = (x^* \vee y) \oplus z = y \oplus z$ and $[x \odot (y \oplus z)] \oplus x^* = x^* \vee (y \oplus z) = y \oplus z$. Therefore $[(x \odot y) \oplus z] \oplus x^* = [x \odot (y \oplus z)] \oplus x^* \neq \mathbf{1}$. The condition follows now from Proposition 67. □

In Section 1.4 we defined BL-homomorphisms, etc. as certain kind of mappings between BL-algebras. If, in particular, we consider MV-algebras we talk about *MV-homomorphisms*, etc. For an MV-homomorphism h we have

$$h(x \oplus y) = h(x^* \to y) = h(x)^* \to h(y) = h(x) \oplus h(y).$$

As was the case in BL-algebra theory, we say that a non-void subset K of an MV-algebra L is a *subalgebra* of L if K is closed with respect to the MV-operations. $[0,1] \cap \mathcal{Q}$, for example, is closed with respect to the Łukasiewicz MV-algebra operations and is therefore a subalgebra of the Łukasiewicz structure. This subalgebra is an incomplete MV-algebra.

Example 1 Let $f : [0,1] \searrow [0,1]$ be a strictly increasing and continuous function, $f(0) = 0$, $f(1) = 1$. We have seen in Exercise 2.1.13 that the structure $W^f = \langle [0,1], \mapsto, ^\star, 1\rangle$, where $x \mapsto y = f^{-1}(1 - f(x) + f(y))$ if $y < x$ and 1 otherwise and $x^\star = f^{-1}(1 - f(x))$ is a Wajsberg algebra. Moreover, W^f is isomorphic to the Łukasiewicz structure $L = \langle [0,1], \to, ^*, 1\rangle$. Indeed, the function $f : [0,1] \searrow [0,1]$ is a mapping $f : W^f \searrow L$ and, for any $x \in [0,1]$, $f(x^\star) = f(f^{-1}(1 - f(x))) = 1 - f(x) = f(x)^*$. Let $x, y \in [0,1]$,

$x \leq y$. Then $f(x \mapsto y) = f(1) = 1 = f(x) \to f(y)$. If $y < x$, then $f(x \mapsto y)$ $= f(f^{-1}(1 - f(x) + f(y))) = 1 - f(x) + f(y) = f(x) \to f(y)$. Thus, f is an MV-homomorphism. Since f is one-to-one, it is an MV-isomorphism.

Thus e.g. functions $f(x) = x^a$, $0 < a$, generate $2^{\aleph_0}$ disjoint Wajsberg algebras on the unit interval which all are complete as the Łukasiewz structure is complete. As we have seen, the algebras $S(m)$ are finite subalgebras of Łukasiewicz structure. Now we introduce a function $d(x, y)$, called *distance function* on an MV-algebra L. This function is defined, for each $x, y \in L$, as follows:

$$d(x, y) = (x^* \odot y) \oplus (y^* \odot x).$$

For the distance function d we have

Proposition 69 *Let L be an MV-algebra. For each $x, y, z, u, v \in L$,*
(i) $d(x, x) = \mathbf{0}$, $d(x, y) = d(y, x)$, $d(x, y) = d(x^*, y^*)$, $d(x, \mathbf{0}) = x$ *and* $d(x, \mathbf{1}) = x^*$,
(ii) *if* $d(x, y) = \mathbf{0}$, *then* $x = y$,
(iii) $d(x, z) \leq d(x, y) \oplus d(y, z)$,
(iv) $d(x \oplus u, y \oplus v) \leq d(x, y) \oplus d(u, v)$,
(v) $d(x \odot u, y \odot v) \leq d(x, y) \oplus d(u, v)$.

Proof. (i) is obvious from the definition. If $d(x, y) = (x^* \odot y) \oplus (y^* \odot x) = \mathbf{0}$, then $(x \oplus y^*) \odot (y \oplus x^*) = \mathbf{1}$, thus $\mathbf{1} = (x \to y) \odot (y \to x) \leq (x \to y)$, $(y \to x)$, therefore $x \leq y$, $y \leq x$, so (ii) holds. In order to prove (iii), we have to show that

$$(x \odot z^*) \oplus (x^* \odot z) \leq (x^* \odot y) \oplus (x \odot y^*) \oplus (y^* \odot z) \oplus (z^* \odot y).$$

By (2.22), (2.36) and (2.20), this is equivalent to

$$(y \to x) \odot (x \to y) \odot (z \to y) \odot (y \to z) \leq (x \to z) \odot (z \to x),$$

which holds as $(x \to y) \odot (y \to z) \leq (x \to z)$ and $(z \to y) \odot (y \to x) \leq (z \to x)$. As for condition (iv), we first note, by setting

$$\mathrm{A} = [x \oplus u \oplus (y^* \odot v^*)] \oplus (x^* \odot y) \oplus (u^* \odot v)$$

that

$$\begin{aligned} \mathrm{A} &= [(x^* \odot y) \oplus x] \oplus (y^* \odot v^*) \oplus [u \oplus (u^* \odot v)] \\ &= [(y^* \odot x) \oplus y] \oplus (y^* \odot v^*) \oplus [v \oplus (v^* \odot u)] \\ &= (y^* \odot x^*) \oplus (v^* \odot u) \oplus v \oplus [y \oplus (y^* \odot v^*] \\ &= (y^* \odot x^*) \oplus (v^* \odot u) \oplus v \oplus [v^* \oplus (y \odot v)] \\ &= \mathbf{1}. \end{aligned}$$

Therefore $[x \oplus u \oplus (y^* \odot v^*)]^* \leq (x^* \odot y) \oplus (u^* \odot v)$, thus

$$(x \oplus u)^* \odot (y \oplus v) \leq (x^* \odot y) \oplus (u^* \odot v).$$

Similarly, we derive

$$(x \oplus u) \odot (y \oplus v)^* \leq (y^* \odot x) \oplus (v^* \odot u).$$

Putting these last two results together we have
$[(x \oplus u)^* \odot (y \oplus v)] \oplus [(x \oplus u) \odot (y \oplus v)^*]$

$$\leq [(x^* \odot y) \oplus (u^* \odot v)] \oplus [(y^* \odot x) \oplus (v^* \odot u)],$$

which is essentially condition (iv). Finally, we reason that

$$\begin{aligned} d(x \odot u, y \odot v) &= d([x^* \oplus u^*]^*, [y^* \oplus v^*]^*) \\ &= d(x^* \oplus u^*, y^* \oplus v^*) \\ &\leq d(x^*, y^*) \oplus d(u^*, v^*) \\ &= d(x, y) \oplus d(u, v). \end{aligned}$$

The proposition is proved. □

Proposition 70 *In an MV-algebra L, for each $x, y \in L$, if $x \leq y$, then $d(x, y) = x^* \odot y$ and $x \oplus d(x, y) = y$.*

Proof. If $x \leq y$, then $x \odot y^* = \mathbf{0}$. Thus $d(x, y) = x^* \odot y$. Also, if $x \leq y$, then $y = y \vee x = x \oplus x^* \odot y = x \oplus d(x, y)$. □

Proposition 71 *In an MV-algebra L, for each $x, y \in L$,*
(i) *if $ny \leq x \leq (n+1)y$, then $d(x, ny) \leq y$ and $d(x, (x+1)y) \leq y$,*
(ii) *if $ny \leq x < (n+1)y$, then $d(x, ny) < y$,*
(iii) *if $ny < x \leq (n+1)y$, then $d(x, (x+1)y) < y$.*

Proof. If $ny \leq x \leq (n+1)y$, then $(ny) \odot x^* = \mathbf{0}$ and $x \odot (y^*)^{n+1} = \mathbf{0}$. Also $d(x, ny) = x \odot (y^*)^n$ and $d(x, (n+1)y) = x^* \odot [(n+1)y]$. From $x \odot (y^*)^{n+1} = \mathbf{0}$, we obtain $(x \odot (y^*)^n) \odot y^* = \mathbf{0}$, thus $d(x, ny) \leq y$. From $(ny) \odot x^* = \mathbf{0}$ and Proposition 59, we see that $d(x, (x+1)y) = x^* \odot (y \oplus ny) \leq y \oplus [x^* \odot (ny)] = y$. Hence (i) is proved. If $d(x, ny) = y$, then $x \odot (y^*)^n = y$ and $x = x \oplus \mathbf{0} = x \oplus (ny) \odot x^* = ny \oplus (y^*)^n \odot x = ny \oplus y = (n+1)y$. This proves (ii). If $d(x, (x+1)y) = y$, then $x^* \odot [(n+1)y] = y$ and $x^* = x^* \oplus \mathbf{0} = x^* \oplus x \odot (y^*)^{n+1} = (y^*)^{n+1} \oplus x^* \odot [(n+1)y] = (y^*)^{n+1} \oplus y = (y^*)^n \odot y^* \oplus y = (y^*)^n \oplus y \odot (ny) \geq (y^*)^n$. Therefore $x \leq ny$. This proves (iii). The proof is complete. □

Proposition 72 *For each element y of an MV-chain L, if $O(y) = m < \infty$, then $d([ny]^*, (m-n)y) < y$ for each $n \leq m$.*

Proof. The claim is clearly true if $n = m$. Therefore let us assume $n < m$. In this case $(m-n)y \oplus ny = \mathbf{1}$, hence $(ny)^* \leq (m-n)y$. On the other hand, if $(ny)^* \leq (m-n-1)y$, then $\mathbf{1} = (m-n-1)y \oplus ny = (m-1)y$ which contradicts $O(y) = m$. Thus we have $(m-n-1)y < (ny)^* \leq (m-n)y$. The claim now follows from Proposition 71 (iii).□

Definition 16 *An element $x \neq \mathbf{0}$ of an MV-algebra L is an* atom *if, whenever $\mathbf{0} \leq y \leq x$, then either $y = \mathbf{0}$ or $y = x$.*

It is an exercise to show that if x is an atom then x^* is co-atom. By Theorem 7 we have

Theorem 14 *If an MV-algebra contains an atom y and $O(y) = m$, then L is isomorphic to $S(m)$.*

Now we consider locally finite MV-algebras with no atoms. By Theorem 14, they are not finite. First we write

Proposition 73 *If L is an MV-chain and $x \leq y$, then either $x \oplus x \leq y$ or $d(x,y) \oplus d(x,y) \leq y$.*

Proof. Assume not $x \oplus x \leq y$. Since L is linearly ordered, $y \leq x \oplus x$ and $\mathbf{1} = y^* \oplus x \oplus x = (y \odot x^*)^* \oplus x$. Hence $(y \odot x^*) \leq x$. By $x \leq y$ we have $y^* \odot x = \mathbf{0}$. Thus $d(x,y) = x^* \odot y \leq x$. By Proposition 70 we see that $d(x,y) \oplus d(x,y) \leq x \oplus d(x,y) = y$. The proof is complete. □

Proposition 74 *If L is an MV-chain and contains no atoms then, for each $x \neq \mathbf{0}$, and for any $n \geq 1$, there exists a $y \neq \mathbf{0}$ for which $ny \leq x$.*

Proof. By induction on n. The claim is clearly true for $n = 1$. Assume that it holds for n and x, y are such that $y \neq \mathbf{0}$ and $ny \leq x$. Since L contains no atoms, there exists a z such that $\mathbf{0} < z < y$. By letting $w = z$ or $w = d(z,y)$, we see from Proposition 73 that $w \neq \mathbf{0}$ and $w \oplus w \leq y$. It is now easy to see that $(n+1)w \leq (2n)w \leq ny \leq x$ and w is the desired element. □

Proposition 75 *If L is a locally finite MV-algebra containing no atoms, then L is* densely ordered, *i.e. for any two elements $x, y \in L$, $x < y$, there exists an element $z \in L$ such that $x < z < y$.*

Proof. Suppose that $x < y$, then, by Proposition 70, $x \oplus d(x,y) = y$ where $d(x,y) \neq \mathbf{0}$. If $d(x,y) = \mathbf{1}$, then, again by Proposition 70, $x^* \odot y = \mathbf{1}$, thus $x = \mathbf{0}$, $y = \mathbf{1}$. In this case by our hypothesis, clearly there will exist an element $z \in L$ such that $x < z < y$. Thus, let us assume $\mathbf{0} < d(x,y) < \mathbf{1}$. Since L is locally finite it is totally ordered and by Proposition 74, there exists an element w such that $\mathbf{0} \neq w \leq d(x,y)$ and $w \oplus w \leq d(x,y)$. Since L is locally finite, $w < d(x,y)$, for otherwise $d(x,y) = d(x,y) \oplus d(x,y)$ and $O(d(x,y)) = \infty$. Thus we have that $\mathbf{0} < w < d(x,y)$. Consider now the element $z = x \oplus w$. Clearly $x \leq z \leq y$. Now, if $z = y$, then $x \oplus w \leq x \oplus w \oplus w \leq x \oplus d(x,y) = y \leq x \oplus w$. Hence, $x \oplus w = x \oplus w \oplus w$. If $y = \mathbf{1}$, then $d(x,y) = x^*$, $w \leq x^*$, $w \oplus w \leq x^*$ and, by (2.78), $w \oplus w = w$. This, of course is a contradiction to the fact that $O(w) < \infty$. If $y \neq \mathbf{1}$, then $x \oplus w = x \oplus w \oplus w \neq \mathbf{1}$, and, by Proposition 67 again we arrive at the contradiction $w \oplus w = w$. Thus we see that $z < y$. Now, if $x = z$ then $z = x \oplus \mathbf{0} = x \oplus w \neq \mathbf{1}$, and, by Proposition 67, we obtain the contradiction $w = \mathbf{0}$. Thus $x < z$. The proposition has been proved. □

The following consideration will establish a connection between MV-chains and ordered additive abelian groups. First we introduce the following

Definition 17 *An algebraic system $\langle G, \leq, +, \mathbf{0} \rangle$ is called* additive ordered abelian group *if $\leq$ is a total order on the non-void set G, the binary operation $+$ is associative and commutative, for each $x \in G$, $x + \mathbf{0} = x$ and there exists an element $-x \in G$, called* additive inverse *of x, such that $x + (-x) = \mathbf{0}$, and for each $x, y, a, b \in G$,*

$$\text{if } a \leq b, \text{ then } x + a + y \leq x + b + y.$$

An element $c \in G$ such that $c > \mathbf{0}$ is called positive element *of G and a set $G[c] = \{x \in G \mid \mathbf{0} \leq x \leq c\}$ is called* positive segment *of G. The element $\mathbf{0}$ is called* identity.

We write simply $x - y$ instead of $x + (-y)$, and as usual we let the underlying set G refer to an additive ordered abelian group $\langle G, \leq, +, \mathbf{0} \rangle$. By taking $G = \mathcal{R}$ or $\mathcal{Q}$, $\leq$ the natural ordering of real numbers, and 0 for the identity we obtain examples of additive ordered abelian groups.

It is easy to see that a totally ordered positive segment $G[c]$ generates an MV-chain when we define $x \oplus y = \min\{c, x + y\}$, $x^* = c - x$, $x \odot y = (x^* \oplus y^*)^*$, $\mathbf{0}$ is the zero and c is the unit. We show that also the converse holds; any MV-chain generates an additive ordered abelian group. Assume L is an MV-algebra. Let L^* be the set of all ordered pairs (m, x) where m is an integer and $x \in L$. On the set L^* we define the following:

$$\begin{aligned}
(m + 1, \mathbf{0}) &= (m, \mathbf{1}), \\
(m, x) \uplus (n, y) &= (m + n, x \oplus y), \text{ if } x \oplus y < \mathbf{1}, \\
(m, x) \uplus (n, y) &= (m + n + 1, x \odot y), \text{ if } x \oplus y = \mathbf{1}, \\
\ominus(m, x) &= (-m - 1, x^*),
\end{aligned}$$

moreover $(m, x) \leq (n, y)$ iff either $m < n$ or $m = n$ and $x \leq y$. Then we have

Proposition 76 *Let L be an MV-chain. Then $\langle L^*, \leq, \uplus, (0, \mathbf{0}) \rangle$ is an additive ordered abelian group.*

Proof. First we prove that L^* is totally ordered with respect to the relation $\leq$. Clearly $(m, x) \leq (m, x)$, and $(m, x) \leq (n, y)$, $(n, y) \leq (m, x)$ is possible only if $m = n$ and $x \leq y$, $y \leq x$, thus $(m, x) = (n, y)$. Assume $(m, x) \leq (n, y)$, $(n, y) \leq (k, z)$. If $m < n$, then $m < k$, thus $(m, x) \leq (k, z)$. If $m = n$, $x \leq y$ and $n < k$, then $m < k$, thus again $(m, x) \leq (k, z)$, and if $m = n$,

$x \leq y$, $n = k$, $y \leq z$, then $m = k$, $x \leq z$, and therefore $(m, x) \leq (k, z)$. We conclude that $L^\star$ is a partially ordered set. Since L is an MV-chain and $\mathcal{Z}$ is totally ordered, so is $L^\star$. To check that the definitions of $\uplus$ and $\ominus$ as given above is consistent with respect to the equality $(m+1, \mathbf{0}) = (m, \mathbf{1})$, we first reason that

$$\begin{aligned} \ominus(m+1, \mathbf{0}) &= (-m-1-1, \mathbf{0}^*) \\ &= (-m-2, \mathbf{1}) \\ &= (-m-1, \mathbf{0}) \\ &= (-m-1, \mathbf{1}^*) \\ &= \ominus(m, \mathbf{1}). \end{aligned}$$

Assume $y < \mathbf{1}$. Then

$$\begin{aligned} (m+1, \mathbf{0}) \uplus (n, y) &= (m+1+n, \mathbf{0} \oplus y) \\ &= (m+1+n, y) \\ &= (m+1+n, \mathbf{1} \odot y) \\ &= (m, \mathbf{1}) \uplus (n, y), \end{aligned}$$

in particular, if $y = \mathbf{1}$, then

$$\begin{aligned} (m, \mathbf{1}) \uplus (n, \mathbf{1}) &= (m+1+n, \mathbf{1} \oplus \mathbf{1}) \\ &= (m+1+n, \mathbf{1}) \\ &= (m+n+2, \mathbf{0}) \\ &= (m+1, \mathbf{0}) \uplus (n+1, \mathbf{0}) \\ &= (m+1, \mathbf{0}) \uplus (n, \mathbf{1}). \end{aligned}$$

It is also clear that the operation $\uplus$ is commutative and

$$(m, x) \uplus [\ominus(m, x)] = (m, x) \uplus (-m-1, x^*) = (m + (-m-1) + 1, x \odot x^*) = (0, \mathbf{0}),$$

thus, $\ominus(m, x)$ is the additive inverse of (m, x). Now we prove that $\uplus$ is associative. Therefore, let three elements (m, x), (n, y), (q, z) of $L^\star$ be given. We wish to show that

$$(m, x) \uplus [(n, y) \uplus (q, z)] = [(m, x) \uplus (n, y)] \uplus (q, z). \tag{2.95}$$

We proceed by cases.

Case 1. $x \oplus y \oplus z < \mathbf{1}$. It is clear that $x \oplus y < \mathbf{1}$ and $y \oplus z < \mathbf{1}$, therefore (2.95) becomes

$$(m + (n+q), x \oplus (y \oplus z)) = ((m+n) + q, (x \oplus y) \oplus z),$$

which certainly holds.

Case 2. $x \oplus y \oplus z = \mathbf{1}$. There are now four subcases.

Case 2a. $x \oplus y < \mathbf{1}$ and $y \oplus z < \mathbf{1}$. In this case (2.95) becomes

$$(m+n+q+1, x \odot (y \oplus z)) = (m+n+q+1, (x \oplus y) \odot z). \tag{2.96}$$

Suppose $x \oplus z = \mathbf{1}$. Then $x^* \leq z$ and $z^* \leq x$ and, by Proposition 68, $x \odot (y \oplus z) = y \oplus (x \odot z) = (x \oplus y) \odot z$ which proves (2.96). Suppose now

$$x \oplus z < \mathbf{1}. \tag{2.97}$$

Since $(x \oplus y)^* \leq z$,

$$z = z \vee (x \oplus y)^* = (x \oplus y) \odot z \oplus (x \oplus y)^*$$

and

$$\begin{aligned} z \oplus x &= (x \oplus y) \odot z \oplus (x \oplus y)^* \oplus x \\ &= (x \oplus y) \odot z \oplus (x^* \odot y^*) \oplus x \\ &= (x \oplus y) \odot z \oplus (x \odot y) \oplus y^*. \end{aligned}$$

Since $x \oplus y < \mathbf{1}$, we have that $x^* \not\leq y$, hence $y \leq x^*$, whence $x \odot y = \mathbf{0}$, therefore

$$z \oplus x = (x \oplus y) \odot z \oplus y^*. \tag{2.98}$$

Similarly, as $(y \oplus z)^* \leq x$, we obtain

$$\begin{aligned} z \oplus x &= z \oplus (y \oplus z) \odot x \oplus (y \oplus z)^* \\ &= x \odot (y \oplus z) \oplus z \oplus (y^* \odot z^*) \\ &= x \odot (y \oplus z) \oplus (y \odot z) \oplus y^* \end{aligned}$$

and, by using the fact $y \oplus z < \mathbf{1}$, we have $y \odot z = \mathbf{0}$, hence

$$z \oplus x = x \odot (y \oplus z) \oplus y^*. \tag{2.99}$$

(2.97), (2.98) and (2.99) enable us to cancel y^* and obtain (2.96).

Case 2b. $x \oplus y < \mathbf{1}$ and $y \oplus z = \mathbf{1}$. In this case the right hand side of (2.95) becomes

$$(m + n + q + 1, (x \oplus y) \odot z), \tag{2.100}$$

and the left hand side of (2.95) becomes

$$(m, x) \uplus (n + q + 1, y \odot z). \tag{2.101}$$

Since $x \oplus y < \mathbf{1}$, hence $x \oplus (y \odot z) < \mathbf{1}$, therefore (2.101) becomes

$$(m + n + q + 1, x \oplus (y \odot z)). \tag{2.102}$$

Using Proposition 68, we see easily that

$$(x \oplus y) \odot z = x \oplus (y \odot z),$$

hence the equality of (2.100) and (2.102) is assured.

Case 2c. $x \oplus y = \mathbf{1}$ and $y \oplus z < \mathbf{1}$. The argument for this case is analogous to that of Case 2b.

Case 2d. $x \oplus y = \mathbf{1}$ and $y \oplus z = \mathbf{1}$. In this case the right hand side of (2.95) becomes

$$(m+n+1, x \odot y) \uplus (q, z), \tag{2.103}$$

and the left hand side of (2.95) becomes

$$(m, x) \uplus (n+q+1, y \odot z)). \tag{2.104}$$

We consider two more subcases.

Case 2d(i). $(x \odot y) \oplus z = \mathbf{1}$. In this case we show that $x \oplus (y \odot z) = \mathbf{1}$. We have that $x^* \oplus y^* \leq z$ and $x^* \leq y$, thus

$$x^* = x^* \wedge y = (x^* \oplus y^*) \odot y \leq z \odot y. \tag{2.105}$$

(2.105) of course implies $x \oplus (y \odot z) = \mathbf{1}$, hence both (2.103) and (2.104) are equal to $(m+n+q+2, x \odot y \odot z)$ which proves (2.95).

Case 2d(ii). $(x \odot y) \oplus z < \mathbf{1}$. In this case by considering the argument in Case 2d(i) and symmetry, we also have $x \oplus (y \odot z) < \mathbf{1}$. Hence (2.103) becomes

$$(m+n+q+1, (x \odot y) \oplus z),$$

and (2.104) becomes

$$(m+n+q+1, x \oplus (y \odot z).$$

We have to show that under these conditions,

$$(x \odot y) \oplus z = x \oplus (y \odot z). \tag{2.106}$$

Since $(x \odot y) \oplus z < \mathbf{1}$, $x \oplus (y \odot z) < \mathbf{1}$, $y^* \leq z$ and $y^* \leq x$ we have, by Proposition 68,

$$(x \odot y \oplus z) \odot y = (x \odot y) \oplus (z \odot y),$$
$$(x \oplus y \odot z) \odot y = (x \odot y) \oplus (z \odot y)$$

and therefore

$$(x \odot y \oplus z) \odot y = (x \oplus y \odot z) \odot y. \tag{2.107}$$

Adding y^* to both sides of (2.107), we get by using the commutativity of $\vee$,

$$y^* \odot z^* \odot (x^* \oplus y^*) \oplus (x \odot y) \oplus z = y^* \odot x^* \odot (y^* \oplus z^*) \oplus x \oplus (y \odot z). \tag{2.108}$$

But since $x \oplus y = y \oplus z = \mathbf{1}$, $x^* \odot y^* = y^* \odot z^* = \mathbf{0}$, therefore (2.108) leads to the desired equality (2.106). Therefore $\uplus$ is associative. It now remains to verify that if (m,x), (n,y), (k,z), $(l,w) \in L^\star$ and $(m,x) \leq (n,y)$, then $(k,z) \uplus (m,x) \uplus (l,w) \leq (k,z) \uplus (n,y) \uplus (l,w)$. This task we leave, however, as an exercise for the reader. □

Theorem 15 *Let L be an MV-chain. Then $L^\star[(0,\mathbf{1})]$ is isomorphic with L, furthermore, the element $(0,\mathbf{1})$ in $L^\star$ has the property that for each $(m,x) \in L^\star$, there is an integer n such that $n(0,\mathbf{1}) \leq (m,x) \leq (n+1)(0,\mathbf{1})$. On the other hand, if G is an additive ordered abelian group and c is a positive element of G such that for each $x \in G$, there exists an n such that $nc \leq x < (n+1)c$, then $G[c]^\star$ is isomorphic with G.*

Proof. First we realize that $L^\star[(0,\mathbf{1})] = \{(0,x) \mid x \in L\}$ and $n(0,\mathbf{1}) = (n-1,\mathbf{1})$ for any integer n. Thus, for any $(m,x) \in L^\star$, $(m-1,\mathbf{1}) \leq (m,x) \leq (m,\mathbf{1})$. To establish an MV-isomorphism between $L^\star[(0,\mathbf{1})]$ and L we need the following observations.

Let $x, y \in L$, $x \oplus y < \mathbf{1}$. Then, in the MV-algebra $L^\star[(0,\mathbf{1})]$,

$$\begin{aligned}(0,x) \oplus (0,y) &= \min\{(0,\mathbf{1}),(0,x) \uplus (0,y)\}\\ &= \min\{(0,\mathbf{1}),(0,x \oplus y)\}\\ &= (0,x \oplus y),\end{aligned}$$

and if $x \oplus y = \mathbf{1}$, then again

$$(0,x) \oplus (0,y) = \min\{(0,\mathbf{1}),(1,x \odot y)\} = (0,\mathbf{1}) = (0,x \oplus y).$$

If $x \neq \mathbf{0}$ then $x^* \neq \mathbf{1}$ and in the MV-algebra $L^\star[(0,\mathbf{1})]$,

$$\begin{aligned}(0,x)^* &= (0,\mathbf{1}) \uplus \ominus (0,x)\\ &= (0,\mathbf{1}) \uplus (-1,x^*)\\ &= (1,\mathbf{0}) \uplus (-1,x^*)\\ &= (0,x^*).\end{aligned}$$

Similarly, we see that $(0,\mathbf{0})^* = (0,\mathbf{0}^*)$. Now we define a function $f : L^\star[(0,\mathbf{1})] \searrow \mathbf{L}$ by $f((0,x)) = x$. Clearly, $f((0,x)) = f((0,y))$ implies $(0,x) = (0,y)$ and f is one-to-one. f is also an MV-homomorphism, indeed $f((0,\mathbf{0})) = \mathbf{0}$,

$$f((0,x) \oplus (0,y)) = f((0,x \oplus y)) = x \oplus y = f((0,x)) \oplus f((0,y)),$$

and

$$f((0,x)^*) = f((0,x^*)) = x^* = [f((0,x))]^*.$$

Therefore

$$\begin{aligned}f((0,x) \to (0,y)) &= f((0,x)^* \oplus (0,y))\\ &= f((0,x))^* \oplus f((0,y))\\ &= f((0,x)) \to f((0,y)).\end{aligned}$$

We conclude that $L^\star[(0,\mathbf{1})]$ is isomorphic with L. For the second part of the theorem we shall exhibit a *group isomorphism* h of G on $G[c]^*$ in the following way. For each $x \in G$, there exists an integer n_x such that $n_x c \leq x < (n_x+1)c$. The function h is defined as follows

$$h(x) = (n_x, x - n_x c).$$

First we reason that n_x is unique as if $nc \leq x < (n+1)c$, $mc \leq x < (m+1)c$ and $m < n$, then $(m+1) \leq n$ and $(m+1)c \leq nc$, thus $nc \leq x < (m+1)c \leq nc$, a contradiction. Similarly, if $n < m$. Thus, $m = n$. Also $n_x c \leq x < (n_x + 1)c$ iff $\mathbf{0} \leq x - n_x c < c$, hence, $(x - n_x c) \in G[c]$, therefore h is well-defined. To verify that h is a group homomorphism, we realize that $n_\mathbf{0} = 0$, thus $h(\mathbf{0}) = (n_\mathbf{0}, \mathbf{0} - n_\mathbf{0}\mathbf{0}) = (0, \mathbf{0})$. Moreover,

$$(n_x + n_y)c \leq x + y < [(n_x + n_y + 1) + 1]c.$$

Thus,

$$\begin{array}{rll} x + y - (n_x + n_y)c < c & \text{iff} & x + y < (n_x + n_y + 1)c \\ & \text{iff} & (n_x + n_y)c \leq x + y < (n_x + n_y + 1)c \\ & \text{iff} & n_x + n_y = n_{x+y}, \end{array}$$

and

$$\begin{array}{c} c \leq x + y - (n_x + n_y)c \ \text{ iff} \\ (n_x + n_y + 1)c \leq x + y < [(n_x + n_y + 1) + 1]c \ \text{ iff} \\ n_x + n_y + 1 = n_{x+y}. \end{array}$$

Therefore, in the MV-algebra $G[c]$,

$$\begin{array}{rcl} (x - n_x c) \oplus (y - n_y c) & = & \min\{c, (x - n_x c) + (y - n_y c)\} \\ & = & \left\{ \begin{array}{ll} x + y - n_{x+y}c, & \text{if } x + y - (n_x + n_y)c < c \\ c, & \text{elsewhere.} \end{array} \right. \end{array}$$

Hence, in the additive ordered abelian group $G[c]^\star$,

$$\begin{array}{l} (n_x, x - n_x c) \uplus (n_y, y - n_y c) \\ = \left\{ \begin{array}{ll} (n_x + n_y, x + y - [n_x + n_y]c), & \text{if } x + y - (n_x + n_y)c < c \\ (n_x + n_y + 1, c - \min\{c, c - [x - n_x c] + c - [y - n_y c]\}), & \text{elsewhere} \end{array} \right. \\ = (n_{x+y}, x + y - [n_{x+y}]c). \end{array}$$

We conclude

$$\begin{array}{rcl} h(x + y) & = & (n_{x+y}, x + y - [n_{x+y}]c) \\ & = & (n_x, x - n_x c)) \uplus (n_y, y - n_y c)) \\ & = & h(x) \uplus h(y) \end{array}$$

and therefore h is a group homomorphism. To see that h one-to-one, we let $(m, z) \in G[c]^*$, and since $(m, c) = (m + 1, \mathbf{0})$, we may assume $\mathbf{0} \leq z < c$. Set $x = z + mc$. Then $\mathbf{0} \leq x - mc < c$ and by the uniqueness of n_x, $m = n_x$. Hence $h(x) = (n_x, x - n_x c) = (m, z)$. It is also easy to see that $h(x) = h(y)$ implies $x = y$. The proof is complete. □

Since the ordered additive abelian group $\langle \mathcal{R}, \leq +, 0 \rangle$ trivially has the property of Theorem 15 and since $\mathcal{R}[1]$ is the Łukasiewicz structure, we now have the desired result

Theorem 16 *Any locally finite MV-algebra is isomorphic with some subalgebra of Łukasiewicz structure.*

It follows by Theorem 16 that any locally finite MV-algebra has at most a continuum number of elements. In the light of Theorem 16 one may also ask if all linear MV-algebras are locally finite and thus isomorphic with some subalgebra of the Łukasiewicz structure. This is not the case. A counter example is a so-called *Chang algebra*, a somewhat special model of MV-algebras obtained by considering the following set C of formal symbols:

$$\begin{gathered} 0, c, c+c, c+c+c, \ldots, \\ 1, 1-c, 1-c-c, 1-c-c-c, \ldots, \end{gathered}$$

For abbreviation, we define $0c = 0$ and $nc = c + \ldots + c + c$ (n-times) and $1 - 0c = 1$ and $1 - nc = 1 - c - \ldots - c - c$ (n-times). In terms of this convention the rules of addition, multiplication, and complementation are written down as follows:

$$x \oplus y = \begin{cases} (m+n)c & \text{if } x = nc, y = mc, \\ 1 & \text{if } x = nc, y = 1 - mc, m \leq n, \\ 1 - (m-n)c & \text{if } x = nc, y = 1 - mc, n < m, \\ 1 & \text{if } x = 1 - mc, y = nc, m \leq n, \\ 1 - (m-n)c & \text{if } x = 1 - mc, y = nc, n < m, \\ 1 & \text{if } x = 1 - nc, y = 1 - mc, \end{cases}$$

$(nc)^* = 1 - nc$, $(1 - nc)^* = nc$ and $x \odot y = (x^* \oplus y^*)^*$. It is routine to check that the system $\langle C, \oplus, \odot, ^*, 0, 1\rangle$ thus defined is an MV-algebra by showing that each of the MV-axioms (2.37)-(2.48) is satisfied. The order relation on this MV-algebra is defined in such a way that $x \leq y$ if, and only if, one of the conditions below is satisfied: (i) $x = nc$ and $y = 1 - mc$. (ii) $x = nc$ and $y = mc$, where $n \leq m$. (iii) $x = 1 - nc$ and $y = 1 - mc$, where $m \leq n$. Obviously, this order is a total order and is not locally finite as $ord(1 - mc) = O(nc) = \infty$.

Remark 17 *Let L, K be two complete MV-algebras. If $h : L \searrow K$ is an MV-homomorphism then, for any subset $\{x_i\}_{i\in\Gamma} \subseteq L$,*

$$\bigvee_{i\in\Gamma} h(x_i) \leq h(\bigvee_{i\in\Gamma} x_i), \tag{2.109}$$

$$h(\bigwedge_{i\in\Gamma} x_i) \leq \bigwedge_{i\in\Gamma} h(x_i). \tag{2.110}$$

Indeed, by (1.85), for any $i \in \Gamma$, $h(\bigwedge_{i\in\Gamma} x_i) \leq h(x_i) \leq h(\bigvee_{i\in\Gamma} x_i)$, which implies (2.109) and (2.110). In general, the sign $\leq$ cannot be replaced by the sign $=$ in (2.109), (2.110) as will be seen in the next

Example 2 Let L be the set of continuous functions on $[0,1]$ to $[0,1]$ with the usual topology. Then L can be equipped with Wajsberg algebra structure by defining, for all $x \in [0,1]$, $(f \to g)(x) = \min\{1, 1-f(x)+g(x)\}$, $f^*(x) = 1-f(x)$, $\mathbf{1}(x) = 1$. Let K be the subalgebra of L consisting of the constant functions. Let $h : L \mapsto K$ be defined as follows: for $f \in L$ and for any $x \in [0,1]$, $h(f)(x) = f(0)$. Thus $h(f) \in K$. h is an MV-homomorphism since $h(f \to g)(x) = (f \to g)(0) = f(0) \to g(0) = h(f)(x) \to h(g)(x) = [h(f) \to h(g)](x)$, and similarly, $h(f^*)(x) = f^*(0) = 1 - f(0) = 1 - h(f)(x) = [h(f)]^*(x)$, $h(\mathbf{1})(x) = \mathbf{1}(x) = 1$. Now, for each natural number n, let f_n be defined by $f_n(x) = 1$ if $\frac{1}{n} \leq x \leq 1$ and $f_n(x) = nx$ if $0 \leq x \leq \frac{1}{n}$. Then $f_n \in L$ and $h(f_n)(x) = f(0) = 0$. Thus

$$\bigvee_{n\in\mathcal{N}} h(f_n) = 0.$$

On the other hand, $\bigvee f_n = \mathbf{1}$. Indeed, $f_n \leq \mathbf{1}$ for any natural number n. Suppose $f \in L$ is such that $f_n \leq f$ for each $n \in \mathcal{N}$. Let $0 < x \leq 1$. Choose an $n \in \mathcal{N}$ so that $\frac{1}{n} \leq x$. Then $f_n(x) = 1$, hence $f(x) = 1$. f is continuous so $f(0) = 1$ as well. Thus $f = \mathbf{1}$. We therefore conclude that

$$h(\bigvee_{n\in\mathcal{N}} f_n) = h(\mathbf{1}) = 1.$$

A counter example concerning (2.110) can be constructed symmetrically. Notice that this MV-algebra is complete and is not linear.

A *fuzzy (sub)set* (of a set X) is, by the original definition of Zadeh, a pair $\langle X, \varphi_X \rangle$, where a mapping $\varphi_X : X \mapsto [0,1]$ is called *membership function.*

Definition 18 *If there is a collection $\mathcal{F}$ of fuzzy sets such that a BL-algebra L and $\mathcal{F}$ are structurally isomorphic, then L is called* representable.

By the results in Section 1.4 we know that any non-degenerate BL-algebra L contains maximal deductive systems M and each such M generates an locally finite MV-algebra L/M. Thus, by the consideration above, there is an MV-monomorphism $k : L/M \mapsto [0,1]$. If $h : L \mapsto L/M$ is the natural BL-homomorphism (i.e. $h(x) = |x|$), then it is an easy exercise to prove that a mapping

$$\varphi_M : L \overset{h}{\mapsto} L/M \overset{k}{\mapsto} [0,1],$$

given by $\varphi_M(x) = k \circ h(x) = k(|x|)$ for any $x \in L$, is a BL-homomorphism. Therefore $\langle L, \varphi_M \rangle$ can be regarded as a fuzzy set. Now set

$$\mathcal{M} = \{M \mid M \text{ is a maximal } ds \text{ of } L\},$$

and for any $x \in L$, define a fuzzy set $\langle M, \widehat{x} \rangle$, where the membership function $\widehat{x} : \mathcal{M} \mapsto [0,1]$ is given via

$$\widehat{x}(M) = \varphi_M(x) \text{ for all } M \in \mathcal{M}.$$

Define, on $\mathcal{F}(L) = \{\widehat{x} \mid x \in L\}$, $\widehat{x} \to \widehat{y} = \widehat{x \to y}$, $\widehat{x}^* = \widehat{x^*}$. Then we have

Theorem 17 *$\langle \mathcal{F}(L), \to, ^*, \widehat{1} \rangle$ is a Wajsberg algebra.*

Proof. First we realize that $\widehat{x} \to \widehat{y} = \widehat{x \to y}$ is a correct definition. Indeed, if $M \in \mathcal{M}$, then $(\widehat{x \to y})(M) = \varphi_M(x \to y) = \varphi_M(x) \to \varphi_M(y) = \widehat{x}(M) \to \widehat{y}(M) = (\widehat{x} \to \widehat{y})(M)$. By a similar argument also $\widehat{x}^* = \widehat{x^*}$ is a correct definition. Now the Wajsberg algebra properties of $\mathcal{F}(L)$ follow by the fact each quotient algebra L/M is a Wajsberg algebra. To prove, for example Wajsberg axiom (2.3), we first realize that in L/M holds $|(x \to y) \to y| = |(y \to x) \to x|$. Thus, for any $x, y \in L$, $M \in \mathcal{M}$, we have

$$\begin{aligned} \varphi_M([x \to y] \to y) &= k(|(x \to y) \to y|) \\ &= k(|(y \to x) \to x|) \\ &= \varphi_M([y \to x] \to x). \end{aligned}$$

Thus,

$$\begin{aligned} ([\widehat{x} \to \widehat{y}] \to \widehat{y})(M) &= (\widehat{[x \to y] \to y})(M) \\ &= \varphi_M([x \to y] \to y) \\ &= \varphi_M([y \to x] \to x) \\ &= (\widehat{[y \to x] \to x})(M) \\ &= ([\widehat{y} \to \widehat{x}] \to \widehat{x})(M) \end{aligned}$$

Hence, for each $\widehat{x}, \widehat{y} \in \mathcal{F}(L)$, $(\widehat{x} \to \widehat{y}) \to \widehat{y} = (\widehat{y} \to \widehat{x}) \to \widehat{x}$. The other Wajsberg axioms can be verified in a similar manner.□

The structure $\mathcal{F}(L)$ is called *Bold algebra of fuzzy sets*. The map $x \mapsto \widehat{x}$ is clearly a BL-epimorphism of L onto $\mathcal{F}(L)$ but is not, in general, an injection.

Theorem 18 *For each BL-algebra L, there is a Bold algebra of fuzzy sets $\mathcal{F}(L)$ and a BL-homomorphism of L onto $\mathcal{F}(L)$. The map $x \mapsto \widehat{x}$ is one-to-one if, and only if $\bigcap\{M \mid M \in \mathcal{M}\} = \{\mathbf{1}\}$.*

Proof. $\widehat{x} = \widehat{y}$ iff for all $M \in \mathcal{M}$ holds $\widehat{x}(M) = \widehat{y}(M)$ iff $\varphi_M(x) = \varphi_M(y)$ iff $|x|_M = |y|_M$ iff $(x \to y) \odot (y \to x) \in M$ iff $x \to y \in M$ and $y \to x \in M$. Thus, condition $\bigcap\{M \mid M \in \mathcal{M}\} = \{\mathbf{1}\}$ means $x \to y = \mathbf{1}$ and $y \to x = \mathbf{1}$, therefore $x = y$ and $x \mapsto \widehat{x}$ is one-to-one. If $\bigcap\{M \mid M \in \mathcal{M}\} \neq \{\mathbf{1}\}$ then there is $y \in \bigcap\{M \mid M \in \mathcal{M}\}$, $y \neq \mathbf{1}$ and so $\widehat{y} \neq \widehat{\mathbf{1}}$. Thus $x \mapsto \widehat{x}$ is not injective.□

A BL-algebra L is called *semi-simple* if $\bigcap\{M \mid M \in \mathcal{M}\} = \{\mathbf{1}\}$. By Theorem 18, a semi-simple BL-algebra L is isomorphic to $\mathcal{F}(L)$ which, by Theorem 17, is an MV-algebra. We therefore have

Proposition 77 *Semi-simple BL-algebras are MV-algebras and representable by fuzzy sets.*

The result above tells us the Łukasiewicz structure has a representation by fuzzy sets while e.g. Brouwer structure or Gaines structure do not posses this kind of representation. It is of importance to realize, too, that Theorem 18 generalizes the famous *Stone representation theorem* on Boolean logic:

> Every Boolean algebra is isomorphic to a subalgebra
> of a direct product of two-element Boolean algebras.

Not all MV-algebras, however, are semi-simple. In the next chapters we will see that what is mostly needed in fuzzy logic and in its applications are complete structures. Fortunately, complete MV-algebras will be semi-simple. To prove this fact will be our next task. As usual, we start by establishing a preliminary result.

Proposition 78 *In any MV-algebra L, if $x \in \bigcap\{M \mid M \in \mathcal{M}\}$ then, for all natural numbers m, n, $x^m \oplus x^n = \mathbf{1}$.*

Proof. Let $x \in \bigcap\{M \mid M \in \mathcal{M}\}$ and we may assume $m \leq n$. An assumption $x \to x^* \in \bigcap\{M \mid M \in \mathcal{M}\}$ leads to a contradiction $x^* \in M$ for all $M \in \mathcal{M}$. Thus, for all $M \in \mathcal{M}$, $x^* \oplus x^* = x \to x^* \notin M$. Since any proper *ds* can be extended to a maximal one, Proposition 31 tells us $ord(x^* \oplus x^*) < \infty$ so, by Proposition 65, $x \oplus x = \mathbf{1}$. Since $x^n \in \bigcap\{M \mid M \in \mathcal{M}\}$, the same argument shows that also $x^n \oplus x^n = \mathbf{1}$. The claim now follows by the fact $x^n \oplus x^n \leq x^m \oplus x^n$.□

Theorem 19 *Let L be an MV-algebra and, for any $x \in L$, $\bigwedge_{m\in\mathcal{N}} x^m \in L$. Then L is semi-simple.*

Proof. Assume $x \in \bigcap_{M\in\mathcal{M}}\{M\}$. Then $\bigwedge_{m\in\mathcal{N}} x^m \in L$ and

$$\bigwedge_{m\in\mathcal{N}} x^m = \bigwedge_{m\in\mathcal{N}}(x \odot x^m) = x \odot \bigwedge_{m\in\mathcal{N}} x^m,$$

where the last equation holds by (2.92). By Exercise 2.1.12, we have $x = x \oplus \bigwedge_{m\in\mathcal{N}} x^m = \bigwedge_{m\in\mathcal{N}}(x \oplus x^m)$, where the last equation holds by (2.87). By Proposition 78, for any natural number m, $x \oplus x^m = \mathbf{1}$. We therefore have $\bigwedge_{m\in\mathcal{N}}(x \oplus x^m) = \mathbf{1}$ and so $x = \mathbf{1}$. Hence L is semi-simple.□

By Theorem 19, any countable complete MV-algebra is representable. We conclude this section by a study on *injective* MV-algebras; they are such MV-algebras L that if A, B are any other MV-algebras and B is a subalgebra of A then any L-valued MV-homomorphism defined on B can be extended to all of A. Injective MV-algebras are known to be complete but not all complete MV-algebras are injective. In fact, an MV-algebra L is injective if, and only if L is complete and ***divisible***, i.e. for any $a \in L$ and any natural number n, there is $b \in L$, called the ***n-divisor*** of a, and such that $nb = a$ and $(a^* \oplus (n-1)b)^* = b$. It is also known that all injective MV-algebras are either isomorphic to Łukasiewicz structure or, more generally, isomorphic to retracts of powers of Łukasiewiz structure. Proving, however, all these facts would take us all too far from elementary algebra; a reader interested in these topics is kindly requested to open the book written by Cignoli, D'Ottaviano and Mundici [8].

We want to keep things simple in this book and give a detailed proof only to the fact that Łukasiewicz structure is injective. Injectivity will play an essential role when we later establish the completeness theorem of fuzzy logic. We start our study by establishing some results on the order of an element x of a locally finite MV-algebra.

Remark 18 *Let L be a locally finite MV-algebra. For any $x \in L$, $x \neq \mathbf{0}, \mathbf{1}$, define $k_1(x) = [(m-1)x]^*$, where $m = O(x)$, and $k_{i+1}(x) = k_1(k_i(x))$ for any natural number i. Then, for any $x, y \neq \mathbf{0}, \mathbf{1}$,*
(i) $k_1(x) \leq x$,
(ii) $k_i(x) \neq \mathbf{0}$.
In particular, if L is Łukasiewicz structure, then
(iii) if $O(x) = O(y)$ and $x < y$ then $k_1(y) < k_1(x)$,
(iv) $k_1(x) = x$ iff $x = \frac{1}{k}$ for some integer $k > 1$.

Proof. Exercise.

Proposition 79 *Consider Łukasiewicz structure. Let $0 < \frac{p}{q} < 1$, p, q natural numbers. Then there is a natural number $j \leq p-1$ such that $k_j(\frac{p}{q})$ $= \frac{1}{s}$, where s divides q.*

Proof. We may assume that p, q do not have common dividers. Let $O(\frac{p}{q}) = m$. If $m \cdot \frac{p}{q} = 1$, then $m \cdot p = q$ and $k_1(\frac{p}{q}) = 1 - (m-1) \cdot \frac{p}{q} = \frac{q - m \cdot p + p}{q} = \frac{p}{q}$. By Remark 18 (iv), $p = 1$ and the Proposition is proved. Otherwise $m \cdot \frac{p}{q}$ > 1 and $m \cdot p = q + r$, where $1 \leq r < p$. In that case $k_1(\frac{p}{q}) = \frac{p-r}{q} = \frac{p_1}{q_1}$, where p_1, q_1 do not have common dividers, $1 \leq p_1 < p, q_1$ and q_1 divides q. Consider now $\frac{p_1}{q_1}$. Then either $p_1 = 1$ or $k_2(\frac{p}{q}) = k_1(\frac{p_1}{q_1}) = \frac{p_2}{q_2}$, where p_2, q_2 do not have common dividers, $1 \leq p_2 < p_1, q_2$ and q_2 divides q_1, thus q_2 divides q, too. Repeat this procedure at most $p-1$ times. Then $k_{p-1}(\frac{p}{q})$ $= \frac{1}{s}$, where s divides q. □

An *MV-formula* (in an MV-algebra L) is obtained in the following finite manner: any $x \in L$ is an MV-formula, and if ϕ, φ are MV-formulas, then $\phi \odot \varphi$, $\phi \oplus \varphi$, φ^* are MV-formulas. In particular, an *MV-term* (in L) $t(x)$ is an MV-formula generated by one element $x \in L$. For example, $(\frac{1}{2} \odot \frac{1}{2}) \oplus \frac{1}{2}^*$ is an MV-term in any MV-algebra in the unit interval, and generated by $\frac{1}{2}$. If L, K are two MV-algebras, $x \in L$, $y \in K$ and $t(x)$ is an MV-term in L, then the MV-term $t(y)$ in K is constructed in the obvious way.

Proposition 80 *Consider Łukasiewicz structure. Let $0 < x < y < 1$. Then there is an MV-term $t(z)$ such that $t(x) \neq t(y)$ and $t(x) \vee t(y) = 1$.*

Proof. If $O(x) \neq O(y) = m = \min\{O(x), O(y)\}$ put $t(z) = mz$. Otherwise let $\frac{p}{q}$ be a rational satisfying $x < \frac{p}{q} < y$. By Remark 18 (iii), $k_1(y) < k_1(\frac{p}{q}) < k_1(x)$. If $O(k_1(y)) \neq O(k_1(x)) = n = \min\{O(k_1(x)), O(k_1(y))\}$ put $t(z) = nk_1(z)$. Otherwise consider $k_2(x) < k_2(\frac{p}{q}) < k_2(y)$ and repeat the previous machinery until one meets the first index j such that $k_j(\frac{p}{q}) = \frac{1}{s}$, where s divides q. By Proposition 79, such an index exists and $\leq p-1$. One will find $k_j(x) < k_j(\frac{p}{q}) < k_j(y)$ if j is even and $k_j(y) < k_j(\frac{p}{q}) < k_j(x)$ if j is odd. Now put $t(z) = sk_j(z)$. □

We are now ready to prove an essential

Theorem 20 *Łukasiewicz structure is an injective MV-algebra.*

Proof. Let A, B be MV-algebras, B is a subalgebra of A, $h : B \mapsto [0,1]$ a Łukasiewicz valued MV-homomorphism defined on B. Since a subset $D = \{b \in B \mid f(b) = 1\}$ is a *ds* of A, it can be extended to a maximal *ds* M and A/M is isomorphic to a subalgebra of Łukasiewicz structure. Thus, there is an MV-monomorphism $m : A/M \mapsto [0,1]$. Let $h : A \mapsto A/M$ be the natural MV-homomorphism. Then $m \circ h : A \mapsto [0,1]$ is trivially an MV-homomorphism defined on all A. Moreover, it has a property

$$f(b) = m \circ h(b) \text{ for all } b \in B. \tag{2.111}$$

Indeed, $m \circ h(a) = m(|a|) = 1$ iff $|a| = |\mathbf{1}|$ (as m is an MV-monomorphism!) iff $a \in M$. Thus, if $f(b) = 1$ then $b \in D \subseteq M$ so $|b| = |\mathbf{1}|$ and therefore also $m \circ h(b) = 1$. On the other hand, if there would be an element $b \in B$ such that $m \circ h(b) = 1$ but $f(b) = k \neq 1$, then $b^m \in M$, where $m = ord(k)$, so that $0 = k^m = f(b)^m = f(b^m)$, hence $1 = 0^* = f(b^m)^* = f([b^m]^*)$, whence $(b^m)^* \in D \subseteq M$; but this simply means $(b^m)^* \odot b^m = \mathbf{0} \in M$, a contradiction. We have now proved

$$\text{for any } b \in B,\ m \circ h(b) = 1 \text{ iff } f(b) = 1,$$

and hence also, for any $b \in B$, $m \circ h(b) = 0$ iff $f(b) = 0$. Assume now there is an element $b \in B$ such that, say, $0 < f(b) < m \circ h(b) < 1$. By Proposition 80, there is an MV-term such that, say, $t(f(b)) = 1$ but $t(m \circ h(b)) \neq 1$. Then we have $t(f(b)) = f(t(b)) = 1$ while $t(m \circ h(b)) = m \circ h(t(b)) \neq 1$, where $t(b) \in B$, a contradiction. Therefore (2.111) holds. The proof is complete.□

2.3 Pseudo-Boolean Algebras

In this section we study pseudo-complemented lattices and pseudo-Boolean algebras which are special kind of residuated lattices and, therefore, are applicable e.g. in the theory of fuzzy relation equations. We have seen that each MV-algebra generates a distributive residuated lattice. Now we will see that each complete MV-algebra generates a pseudo-Boolean algebra. We also study the basic properties of this algebraic structure. We start by setting the following definition.

Definition 19 *A lattice L containing the zero element $\mathbf{0}$ and the unit element $\mathbf{1}$ is a called* pseudo-complemented lattice *if, for any $x, y \in L$, there exists the greatest element $z \in L$ such that $x \wedge z \leq y$. This element is denoted by $x \Rightarrow y$. By definition, for each $x, y \in L$,*

$$x \Rightarrow y = \bigvee\{z \in L \mid x \wedge z \leq y\}, \tag{2.112}$$

or equivalently, for each $x, y, z \in L$,

$$z \leq x \Rightarrow y \text{ iff } x \wedge z \leq y. \tag{2.113}$$

If, in particular, L is distributive we talk about pseudo-Boolean algebra.

Since in any complete lattice L the operation $\wedge$ is associative, commutative, isotone and, if the unit element $\mathbf{1}$ exists in L then, for each $x \in L$, each subset $\{y_i\}_{i\in\Gamma} \subseteq L$, holds $x \wedge \mathbf{1} = x$, $\bigvee_{i\in\Gamma}(x \wedge y_i) = x \wedge \bigvee_{i\in\Gamma} y_i$, we conclude

Proposition 81 *A pseudo-complemented lattice L forms a residuated lattice $\langle L, \leq, \wedge, \vee, \wedge, \Rightarrow, \mathbf{0}, \mathbf{1}\rangle$. Moreover, a complete MV-algebra, when considered as a lattice, generates a pseudo-Boolean algebra and $\langle \wedge, \Rightarrow\rangle$ is an adjoint couple.*

Assume a pseudo-complemented lattice L is a chain and $a, b \in L$. By definition, if $a \leq b$, then $a \Rightarrow b = \mathbf{1}$. Otherwise $a \Rightarrow b = b$.

Definition 20 *Let L be a pseudo-complemented lattice. An element $a^+ \in L$, defined by $a^+ = a \Rightarrow 0$, is the* pseudo-complement *of an element $a \in L$.*

Notice that the pseudo-complement a^+ of an element a can be defined in any such lattice L that $\bigvee\{x \in L \mid a \wedge x = \mathbf{0}\}$ exists in L. In Boolean algebras complement, l-complement and pseudo-complement coincide, while in MV-algebras complement and pseudo-complement are disjoint; in the Łukasiewicz structure, for example, $\frac{1}{2}$ has no l-complement, $\frac{1}{2}^* = \frac{1}{2}$, and $\frac{1}{2}^+ = 0$.

Proposition 82 *In a pseudo-complemented lattice L, for each element x, x_1, x_2, y, y_1, y_2, $z \in L$,*

$$x \Rightarrow y = \mathbf{1} \text{ iff } x \leq y, \tag{2.114}$$

$$x = y \text{ iff } x \Rightarrow y = y \Rightarrow x = \mathbf{1}, \tag{2.115}$$

$$x \Rightarrow x = \mathbf{1}, \tag{2.116}$$

$$x \Rightarrow \mathbf{1} = \mathbf{1}, \tag{2.117}$$

$$\mathbf{1} \Rightarrow y = y, \tag{2.118}$$

$$(x \Rightarrow x) \wedge y = y, \tag{2.119}$$

$$x \wedge (x \Rightarrow y) \leq y, \tag{2.120}$$

$$\textit{if } x_1 \leq x_2, \textit{ then } x_2 \Rightarrow y \leq x_1 \Rightarrow y, \tag{2.121}$$

$$\textit{if } y_1 \leq y_2, \textit{ then } x \Rightarrow y_1 \leq x \Rightarrow y_1, \tag{2.122}$$

$$y \leq (x \Rightarrow y), \tag{2.123}$$

$$x \wedge (x \Rightarrow y) \leq (x \wedge y), \tag{2.124}$$

$$(x \Rightarrow y) \wedge y \leq y, \tag{2.125}$$

$$(x \Rightarrow y) \wedge (x \Rightarrow z) = x \Rightarrow (y \wedge z), \tag{2.126}$$

$$(x \Rightarrow z) \wedge (y \Rightarrow z) = (x \vee y) \Rightarrow z, \tag{2.127}$$

$$x \Rightarrow (y \Rightarrow z) = (x \wedge y) \Rightarrow z = y \Rightarrow (x \Rightarrow z), \tag{2.128}$$

$$z \Rightarrow x \leq [z \Rightarrow (x \Rightarrow y)] \Rightarrow (z \Rightarrow y), \tag{2.129}$$

$$(x \Rightarrow y) \wedge (y \Rightarrow z) \leq (x \Rightarrow z), \tag{2.130}$$

$$x \Rightarrow y \leq (y \Rightarrow z) \Rightarrow (x \Rightarrow z), \tag{2.131}$$

$$x \leq y \Rightarrow (x \wedge y), \tag{2.132}$$

$$x \Rightarrow (y \Rightarrow z) \leq (x \Rightarrow y) \Rightarrow (x \Rightarrow z), \tag{2.133}$$

$$z \wedge [(z \wedge x) \Rightarrow (z \wedge y)] = z \wedge (x \Rightarrow y). \tag{2.134}$$

Proof. (2.114)-(2.123), (2.126)-(2.128), (2.130)-(2.132) are direct consequences from the fact that $\langle \wedge, \Rightarrow \rangle$ is an adjoint couple. (2.125) is trivial. As $x \wedge (x \Rightarrow y) \leq x$ and, by (2.120), $x \wedge (x \Rightarrow y) \leq y$ we have (2.124). Since $z \Rightarrow (x \wedge y) \leq z \Rightarrow y$, we have, by (2.124), that $z \Rightarrow [x \wedge (x \Rightarrow y)] \leq z \Rightarrow y$ and consequently, by (2.126), $(z \Rightarrow x) \wedge [z \Rightarrow (x \Rightarrow y)] \leq z \Rightarrow y$. Thus $(z \Rightarrow x) \leq [z \Rightarrow (x \Rightarrow y)] \Rightarrow (z \Rightarrow y)$, i.e. (2.129). By (1.46),

$$y \Rightarrow z \leq (x \Rightarrow y) \Rightarrow (x \Rightarrow z),$$

which implies $x \Rightarrow (y \Rightarrow z) \leq x \Rightarrow [(x \Rightarrow y) \Rightarrow (x \Rightarrow z)]$. Now

$$\begin{aligned} x \Rightarrow [(x \Rightarrow y) \Rightarrow (x \Rightarrow z)] &= (x \Rightarrow y) \Rightarrow [x \Rightarrow (x \Rightarrow z)] && \text{(by (1.48))} \\ &= (x \Rightarrow y) \Rightarrow [(x \wedge x) \Rightarrow z] && \text{(by (1.47))} \\ &= (x \Rightarrow y) \Rightarrow (x \Rightarrow z). && \text{(by (1.10))} \end{aligned}$$

Therefore (2.133) holds. Since $x \Rightarrow y \leq (z \wedge x) \Rightarrow y$, $x \Rightarrow y \leq \mathbf{1} = (z \wedge x) \Rightarrow z$, we have

$$x \Rightarrow y \leq [(z \wedge x) \Rightarrow y] \wedge [(z \wedge x) \Rightarrow z] = (z \wedge x) \Rightarrow (z \wedge y).$$

Therefore $z \wedge (x \Rightarrow y) \leq z \wedge [(z \wedge x) \Rightarrow (z \wedge y)]$. Conversely,

$$(z \wedge x) \wedge [(z \wedge x) \Rightarrow (z \wedge y)] \leq z \wedge y \leq y,$$

therefore $z \wedge [(z \wedge x) \Rightarrow (z \wedge y)] \leq x \Rightarrow y$, consequently, $z \wedge [(z \wedge x) \Rightarrow (z \wedge y)] \leq z \wedge (x \Rightarrow y)$. Thus, (2.134) holds. □

Since $\langle \wedge, \Rightarrow \rangle$ is an adjoint couple we also have

Proposition 83 *In a pseudo-complemented lattice L, for each x, $y \in L$,*

$$\text{if } x \leq y, \text{ then } y^+ \leq x^+, \tag{2.135}$$

$$\mathbf{0}^+ = \mathbf{1}, \mathbf{1}^+ = \mathbf{0}, \tag{2.136}$$

$$x \wedge x^+ = \mathbf{0}, \tag{2.137}$$

$$(x \wedge x^+)^+ = \mathbf{1}, \tag{2.138}$$

$$x \leq x^{++}, \tag{2.139}$$

$$x^{+++} = x^+, \tag{2.140}$$

$$\mathbf{0} \Rightarrow x = \mathbf{1}. \tag{2.141}$$

Proposition 84 *In a pseudo-Boolean algebra L, for each $x, y \in L$,*

$$(x \vee y)^+ = x^+ \wedge y^+, \tag{2.142}$$

$$x^+ \vee y^+ \leq (x \wedge y)^+, \tag{2.143}$$

$$x^+ \vee y \leq x \Rightarrow y, \tag{2.144}$$

$$x \Rightarrow y \leq y^+ \Rightarrow x^+, \tag{2.145}$$

$$x \Rightarrow y^+ = (x \wedge y)^+, \tag{2.146}$$

$$x \Rightarrow y^+ = y \Rightarrow x^+, \tag{2.147}$$

$$x \Rightarrow y^+ = (x \Rightarrow y^+)^{++}, \tag{2.148}$$

$$(x \Rightarrow y)^{++} \leq x \Rightarrow y^{++}. \tag{2.149}$$

Proof. Exercise. Notice that, in general $x \neq x^{++}$ and $x^+ \vee x \neq \mathbf{1}$, take e.g. the Łukasiewicz structure and $x \neq \mathbf{0}, \mathbf{1}$. Notice also that in a linearly ordered pseudo-complemented lattice L, for any $x, y \in L$,

$$x^+ \vee y^+ = (x \wedge y)^+. \tag{2.150}$$

Indeed, since L is linearly ordered we have $x \wedge y = x$ or y. Assume $x \wedge y = y$. Then $(x \wedge y)^+ = y^+$. Since $y \leq x$, we have, by (2.135), $x^+ \leq y^+$ and $x^+ \vee y^+ = y^+$. Therefore $x^+ \vee y^+ = (x \wedge y)^+$. The case $x \wedge y = x$ is symmetric.

Proposition 85 *In a pseudo-Boolean algebra L, for each x, y, z, $w \in L$,*

$$(x \vee y) \wedge x^+ \leq y, \tag{2.151}$$

$$(x \vee y) \wedge (x \Rightarrow z) \wedge (y \Rightarrow w) \leq z \vee w. \tag{2.152}$$

Proof. By distributivity and (2.137), $(x \vee y) \wedge x^+ = (x \wedge x^+) \vee (y \wedge x^+) = \mathbf{0} \vee (y \wedge x^+) \leq y \wedge x^+$. Thus (2.151) holds. To prove (2.152) we reason first that $x \wedge (x \Rightarrow z) \leq z$, $y \wedge (y \Rightarrow w) \leq w$, therefore

$$[x \wedge (x \Rightarrow z) \wedge (y \Rightarrow w)] \vee [y \wedge (x \Rightarrow z) \wedge (y \Rightarrow w)] \leq z \vee w.$$

By distributivity, $(x \vee y) \wedge (x \Rightarrow z) \wedge (y \Rightarrow w) \leq z \vee w$. Thus (2.152) holds. □

2.4 Exercises

2.1. MV-Algebras and Wajsberg algebras

Exercise 1 Prove (a) Remark 13, (b) Remark 14. Do this by using only results occurring in text before them.

Exercise 2 Verify equations (2.20)-(2.22).

Exercise 3 Verify equations (2.25).

Exercise 4 Define on a Wajsberg algebra L a binary operation $\odot$, for any $x, y \in L$, by $x \odot y = (x \to y^*)^*$. Prove that
(a) $x \odot \mathbf{1} = x$ holds for any $x \in L$,
(b) $x \odot y \leq z$ iff $x \leq y \to z$ holds for any $x, y, z \in L$.

Exercise 5 Prove that equation (2.35) holds in a Wajsberg algebra.

Exercise 6 Verify (2.39) - (2.48), in other words, prove that a Wajsberg algebra generates an MV-algebra.

Exercise 7 Bearing in mind the proof of Proposition 52, verify that MV-axioms (2.39)-(2.48) imply, for any $x, y, z \in L$,
(a) $(x \wedge y)^* = x^* \vee y^*$,
(b) $x \vee (x \wedge y) = x$,
(c) if $x \odot y = \mathbf{1}$, then $x = y = \mathbf{1}$,
(d) if $x \wedge y = \mathbf{1}$, then $x = y = \mathbf{1}$,
(e) if $x \leq y$, then $x \wedge z \leq y \wedge z$,
(f) if $x, y \leq z$, then $x \vee y \leq z$,
(g) if $x \leq y$, then $x \odot z \leq y \odot z$.

Exercise 8 After Theorem 10 we remarked that in such a residuated lattice that (2.79) holds we have l.u.b.$\{x, y\} = (x \to y) \to y$. In such a case prove, for all $x, y \in L$, that also g.l.b$\{x, y\} = x \odot (x \to y) = y \odot (y \to x)$ holds.

Exercise 9 Verify that Boolean algebras are Wajsberg algebras by showing, for all elements x, y, that $(x \to y) \to y = (y \to x) \to x$, where $x \to y = x^* \vee y$.

Exercise 10 Prove that in an MV-algebra L, for any $x \in L$, $x \odot x = x$ iff $x \oplus x = x$.

Exercise 11 Prove that the structures defined in Example 1 of Section 1.3. do not generate, in general, Wajsberg algebra structures, while generalized Lukasiewicz structures generate Wajsberg algebras.

Exercise 12 Prove that in an MV-algebra L, for all $x, y \in L$, the following are equivalent conditions: (i) $x \oplus y = y$, (ii) $x \odot y = x$, (iii) $x^* \vee y = \mathbf{1}$, (iv) $x \wedge y^* = \mathbf{0}$.

Exercise 13 Let $f : [0,1] \to [0,k]$, $0 < k < \infty$, be a strictly increasing and continuous function with $f(0) = 0$, $f(1) = k$. Then f has the inverse function f^{-1}. Prove that we obtain an MV-algebra by defining, for any $x, y \in I$,
$x \oplus y = f^{-1}(f(x) + f(y))$ for $f(x) + f(y) \in [0,k]$ and 1 otherwise,
$x^* = f^{-1}(k - f(x))$, $x \odot y = (x^* \oplus y^*)^*$.

Exercise 14 Letting L denote a general BL-algebra, consider a subset $MV(L) = \{x^* \mid x \in L\}$. Prove that $MV(L)$ is not empty, closed with respect to residuum, and satisfies the Wajsberg algebra axioms.

2.2. Complete MV-algebras

Exercise 1 Prove that $\bigvee_{i \in \Gamma} y_i^* = (\bigwedge_{i \in \Gamma} y_i)^*$ holds in a complete MV-algebra.

Exercise 2 Prove equation (1.17) particularly for MV-algebras.

Exercise 3 Let L be a residuated lattice generated by a complete MV-algebra. Prove that L is continuous.

Exercise 4 Consider a complete residuated lattice defined on the unit interval. For each $a \in [0,1]$, define functions g_a by $g_a(x) = a \to x, x \in [0,1]$. Prove that g_a are continuous functions iff (2.93) holds.

Exercise 5 Let $f(x) = e^x - 1$ in Example 2. Write down the equations for $x \odot y$, $x \oplus y$, $x \to y$ and x^*.

Exercise 6 Let the real unit interval be equipped with a complete MV-algebra structure. For each $a \in [0,1]$, define h_a and k_a by $h_a(x) = a \odot x$, $k_a(x) = a \oplus x$, $x \in [0,1]$. Prove that h_a and k_a are continuous functions.

Exercise 7 Verify the following equations in MV-algebras:
(a) $x \oplus y = (x \vee y) \oplus (x \wedge y)$ and, (b) $x \odot y = (x \vee y) \odot (x \wedge y)$.

Exercise 8 Prove, for all elements x of an MV-algebra L and all $n \geq 0$
(a) $(nx)^* = (x^*)^n$ and, (b) $(x^n)^* = n(x^*)$.

Exercise 9 Show, for all $n \geq 1$, $x, y \in L$, L an MV-algebra, the following:
(a) $(x \vee y)^n = x^n \vee y^n$, (b) $(x \wedge y)^n = x^n \wedge y^n$, (c) $n(x \vee y) = nx \vee ny$ and,
(d) $n(x \wedge y) = nx \wedge ny$.

Exercise 10 If $x \wedge y = \mathbf{0}$ in an MV-algebra prove, for all natural numbers n, that $nx \wedge ny = \mathbf{0}$.

Exercise 11 Prove: if $ord(x \oplus y) < \infty$ then $x \odot y = \mathbf{0}$, where x, y are elements of an MV-algebra.

Exercise 12 Prove in an MV-algebra: if x is an atom then x^* is a co-atom.

Exercise 13 Complete the proof of Proposition 76.

Exercise 14 Let L be a BL-algebra and M a maximal *ds* of L. Show that a mapping $\varphi_M : L \overset{h}{\mapsto} L/M \overset{k}{\mapsto} [0,1]$, where h is the natural BL-homomorphism and m is a BL-monomorphism, is a BL-homomorphism.

Exercise 15 Prove Remark 18.

Exercise 16 Demonstrate that the direct product algebra of two injective MV-algebras is an injective MV-algebra.

Exercise 17 Recalling the definition of divisible algebra on the real unit interval, prove that Łukasiewicz structure is divisible while Gödel structure and Product structure are not divisible.

2.3. Pseudo-Boolean algebras

Exercise 1 Verify (2.142) - (2.149).

Exercise 2 A unary operation η, defined on an MV-algebra L, and such that
(i) $\eta a = a^*$ for any $a \in BoL$, (ii) $a \leq b$ implies $\eta b \leq \eta a$ for any $a, b \in L$, is called *negation*[3]. Prove that pseudo-complement is negation.

[3]Definitions in Exercise 2 are due to Belluce.

Chapter 3

Fuzzy Propositional Logic

Now we start to develop multiple valued logic, that is, logic with more truth values than only 'false' and 'true', or 0 and 1. We assume all the time that the set L of values of truth forms at least a complete lattice. Thus, the set of values of truth is partially ordered and there is the least element $\mathbf{0}$ and the greatest element $\mathbf{1}$ in L corresponding to absolute false and absolute truth, respectively. Usually the set L of values of truth is the unit interval but also other alternatives are possible; we will see that any injective MV-algebra will offer sufficient conditions for the well behavior of multiple valued logic and necessary condition for the well behavior of our logic is that L is a complete MV-algebra.

3.1 Semantics of Fuzzy Propositional Logic

The formalized language of Fuzzy Propositional Logic (Fuzzy Logic, in short) is composed of four kinds of building blocks:
(i) The set of ***propositions*** or ***propositional variables*** is an infinite set $L = \{p_i; i \in \mathcal{N}\}$. Propositions are sometimes denoted by p, q, r, s, t, w, too.
(ii) for any element $a \in L$ there is an *inner truth value* $\mathbf{a}$ in the formalized language of Fuzzy Propositional Logic.
(iii) The logical connectives of Fuzzy Propositional Logic are `imp` (read 'implies') and **and** (read 'and' or 'conjunction').
(iv) There are auxiliary symbols $\},],), (, [, \{$ in the language of Fuzzy Propositional Logic.

Inner truth values and propositional variables are *atomic formulas.*

Definition 21 *The set $\mathcal{F}$ of well-formed formulas is constructed in the following way:*
(i) *atomic formulas are in $\mathcal{F}$,*
(ii) *if α and β are in $\mathcal{F}$ then $(\alpha$ `imp` $\beta)$ and $(\alpha$ **and** $\beta)$ are in $\mathcal{F}$.*

Propositional variables correspond statements like *John is tall, John is very tall, It is raining, It is slightly raining, This power plant is performing well, This power plant is usually performing well,* etc. We do not study the mutual relation of propositions like *John is tall* and *John is very tall,* but take them as atoms. Instead, we tend to think that the fact *This power plant is usually performing well,* for example, can be expressed by associating a truth value close to 1 (but not 1!) with an expression *This power plant is performing well.* In general, we are more interested in the mathematical properties of the language of Fuzzy Propostional Logic as a whole.

At this stage it should be pointed out that the language of Fuzzy Propositional Logic is much simpler than natural language. For example, there is no causal link between formulas α and β in the formula $(\alpha \ \mathtt{imp} \ \beta)$, nor is there no temporal order among them in the formula $(\alpha \ \mathbf{and} \ \beta)$. The statement *If I switch off the light, then there will be dark in this room* is quite correct in natural language and contains an idea of causal link between the statements *I switch off the light* and *There is dark in this room* while the statement *If the Moon is a big cheese then A is equal to A* sounds odd and meaningless in natural language; whatever the Moon is A is equal to itself anyway. In the formalized language of Fuzzy Propositional Logic both statements, however, are well-formed formulas of the form $(\alpha \ \mathtt{imp} \ \beta)$. Similarly, the expression *During the lunch hour Giuseppe went home and made love* has different meaning in natural language than the expression *During the lunch hour Giuseppe made love and went home.* In the formalized language of Fuzzy Propositional Logic this kind of temporal difference between formulas $(\alpha \ \mathbf{and} \ \beta)$ and $(\beta \ \mathbf{and} \ \alpha)$ does not occur. To avoid unnecessary parentheses we may write $(\alpha \ \mathbf{and} \ \beta \ \mathbf{and} \ \gamma)$ instead of $[(\alpha \ \mathbf{and} \ \beta) \ \mathbf{and} \ \gamma]$ or $[\alpha \ \mathbf{and} \ (\beta \ \mathbf{and} \ \gamma)]$. An expression $(\alpha \ \mathtt{imp} \ \beta \ \mathtt{imp} \ \gamma)$ is not, however, a well-formed formula; one should write either $[(\alpha \ \mathtt{imp} \ \beta) \ \mathtt{imp} \ \gamma]$ or $[\alpha \ \mathtt{imp} \ (\beta \ \mathtt{imp} \ \gamma)]$ which are two different formulas.

The inner truth values $\mathbf{a}$, $a \in L$, generalize the falsum sign $\perp$ of classical logic. We follow the idea of intuitionistic logic and abbreviate $(\alpha \ \mathtt{imp} \ \mathbf{0})$ by $(\mathtt{non} - \alpha)$. The logical connective $\mathtt{non}$ (read 'non') is called *negation.*

Giving *semantic interpretation* to a formula $\alpha \in \mathcal{F}$ means we associate a value of truth $v(\alpha) \in L$ to α, in other words, we define *truth value function* $v : \mathcal{F} \searrow L$. Before introducing an exact mathematical definition of truth value function v, let us discuss the properties this function should obey. A very natural assumption is (V1):

$v(\alpha)$ is defined for each well-formed formula $\alpha \in \mathcal{F}$.

Since we are generalizing Classical Propositional Logic and since for all classical truth value functions v holds $v(\perp) = 0$, we set (V2):

For all inner truth values $\mathbf{a}$, $v(\mathbf{a}) = a$.

We adopt *Principle of Truth Functionality* into fuzzy logic, namely, (V3):

For all $\alpha, \beta \in \mathcal{F}$, $v(\alpha \texttt{ imp } \beta)$, $v(\alpha \textbf{ and } \beta)$ *depend only on* $v(\alpha)$, $v(\beta)$.

A crucial question in defining truth value for a formula containing the logical connective imp is *What should the value* $v(\alpha \texttt{ imp } \beta)$ *tell us?* As there is not necessarily any causal link between formulas α and β in a formula $(\alpha \texttt{ imp } \beta)$ we accept *Principle for Implication* (PI):

The degree of truth of a formula $(\alpha \texttt{ imp } \beta)$ *quantifies the degree by which* β *is at least as true as* α.

In other words, (PI) states that the degree $v(\alpha \texttt{ imp } \beta)$ quantifies the degree by which $v(\beta)$ is at least as large as $v(\alpha)$. A direct consequence of (PI) is *Boundary Condition* (BC):

For all $\alpha, \beta \in \mathcal{F}$, $v(\alpha \texttt{ imp } \beta) = \mathbf{1}$ *iff* $v(\alpha) \leq v(\beta)$.

Remark 19 *One of the most important properties of the logical connective* imp *is that* imp *defines a quasi-order on the set of well-formed formulas. Accepting* (PI), *this property holds in Fuzzy Propositional Logic, too.*

Indeed, define, for all $\alpha, \beta \in \mathcal{F}$, $\alpha \preceq \beta$ iff $v(\alpha \texttt{ imp } \beta) = \mathbf{1}$. Then, as $v(\alpha) \leq v(\alpha)$, we have $v(\alpha \texttt{ imp } \alpha) = \mathbf{1}$, thus $\alpha \preceq \alpha$ for any $\alpha \in F$. Moreover, if $\alpha \preceq \beta$ and $\beta \preceq \gamma$, then $v(\alpha \texttt{ imp } \beta) = \mathbf{1}$, $v(\beta \texttt{ imp } \gamma) = \mathbf{1}$, thus $v(\alpha) \leq v(\beta) \leq v(\gamma)$ and so $v(\alpha \texttt{ imp } \gamma) = \mathbf{1}$, hence $\alpha \preceq \gamma$.

Another consequence of (PI) is *Monotonicity of Implication* (MI):

For all $\alpha, \beta, \gamma \in \mathcal{F}$, *if* $v(\alpha) \leq v(\beta)$, *then* $v(\gamma \texttt{ imp } \alpha) \leq v(\gamma \texttt{ imp } \beta)$, $v(\beta \texttt{ imp } \gamma) \leq v(\alpha \texttt{ imp } \gamma)$.

The inner truth value **1** corresponds to tautology in Classical Propositional Logic. It is therefore natural to assume that the degree of truth of a formula $(\alpha \textbf{ and } \mathbf{1})$ coincides with the degree of truth of the formula α, by symbols (V4):

For all $\alpha \in \mathcal{F}$, $v(\alpha \textbf{ and } \mathbf{1}) = v(\alpha)$.

Natural conditions for the degree of truth of formulas containing the connective **and** are also the following (V5):

For all $\alpha, \beta, \gamma \in \mathcal{F}$, *if* $v(\alpha) \leq v(\beta)$, *then* $v(\alpha \textbf{ and } \gamma) \leq v(\beta \textbf{ and } \gamma)$

and (V6):

For all $\alpha, \beta, \gamma \in \mathcal{F}$, $v(\alpha \textbf{ and } (\beta \textbf{ and } \gamma)) = v((\alpha \textbf{ and } \beta) \textbf{ and } \gamma)$.

Finally, the degree of truth of a formula $[(\alpha \textbf{ and } \beta) \texttt{ imp } \gamma]$ should coincide with the degree of truth of the formula $[\alpha \texttt{ imp } (\beta \texttt{ imp } \gamma)]$ as well as with the degree of the truth of the formula $[\beta \texttt{ imp } (\alpha \texttt{ imp } \gamma)]$, i.e. (V7):

For all $\alpha, \beta, \gamma \in \mathcal{F}$,
$$v((\alpha \texttt{ and } \beta) \texttt{ imp } \gamma) = v(\alpha \texttt{ imp } (\beta \texttt{ imp } \gamma)) = v(\beta \texttt{ imp } (\alpha \texttt{ imp } \gamma)).$$

Theorem 21 *Assume there exists a function* $v : \mathcal{F} \searrow L$ *such that all the conditions mentioned above hold. Then the set* L *of values of truth is a residuated lattice.*

Proof. By Principle of Truth Functionality, there are binary operations $\odot$ and $\to$ defined on the complete lattice $\langle L, \leq, \wedge, \vee \rangle$ of values of truth such that $v(\alpha \texttt{ and } \beta) = v(\alpha) \odot v(\beta)$, $v(\alpha \texttt{ imp } \beta) = v(\alpha) \to v(\beta)$. We demonstrate that $\langle L, \leq, \wedge, \vee, \odot, \to \rangle$ is a residuated lattice. By (V2), for any $a \in L$, there is an inner truth value $\mathbf{a} \in \mathcal{F}$ such that $v(\mathbf{a}) = a$, thus $a \odot b$ and $a \to b$ are always defined. Obviously, this definition is unique. By (V5), $\odot$ is isotone and, by (V6), also associative. The operation $\odot$ is commutative as $v(\alpha \texttt{ and } \beta) \leq v(\alpha \texttt{ and } \beta)$ iff $v((\alpha \texttt{ and } \beta) \texttt{ imp } (\alpha \texttt{ and } \beta)) = \mathbf{1}$ iff $v(\alpha \texttt{ imp } [\beta \texttt{ imp } (\alpha \text{ tt and } \beta)] = \mathbf{1}$ iff $v(\beta \texttt{ imp } [\alpha \texttt{ imp } (\alpha \texttt{ and } \beta]) = \mathbf{1}$ iff $v([\beta \texttt{ and } \alpha] \texttt{ imp } [\alpha \texttt{ and } \beta]) = \mathbf{1}$ iff $v(\beta \texttt{ and } \alpha) \leq v(\alpha \texttt{ and } \beta)$. By a similar argument $v(\alpha \texttt{ and } \beta) \leq v(\beta \texttt{ and } \alpha)$. Therefore $v(\alpha \texttt{ and } \beta) = v(\beta \texttt{ and } \alpha)$. Hence, for any $a, b \in L$, $a \odot b = v(\mathbf{a}) \odot v(\mathbf{b}) = v(\mathbf{a} \texttt{ and } \mathbf{b}) = v(\mathbf{b} \texttt{ and } \mathbf{a}) = v(\mathbf{b}) \odot v(\mathbf{a}) = b \odot a$. Since $v(\alpha) = v(\alpha \texttt{ and } \mathbf{1}) = v(\alpha) \odot v(\mathbf{1}) = v(\alpha) \odot \mathbf{1}$, we conclude that $a = a \odot \mathbf{1}$ for any $a \in L$. Finally, since $v(\alpha \texttt{ and } \beta) \leq v(\gamma)$ iff $v((\alpha \texttt{ and } \beta) \texttt{ imp } \gamma) = \mathbf{1}$ iff $v(\alpha \texttt{ imp } (\beta \texttt{ imp } \gamma)) = \mathbf{1}$ iff $v(\alpha) \leq v(\beta \texttt{ imp } \gamma)$, it is easy to see that $\langle \odot, \to \rangle$ is an adjoint couple. □

It is natural to assume that small variation in the value of truth in any argument of a formula of the form $(\alpha \texttt{ imp } \beta)$ should not lead to a large change of value of truth in the whole formula. This requires *Continuity of Implication* (V8):

For all $\alpha, \beta \in \mathcal{F}$, $v(\alpha \texttt{ imp } \beta)$ *is continuous in its arguments.*

Thus, by the consideration in the previous sections, to fulfill this postulate, the set of values of truth must form a complete MV-algebra. In particular, as injective MV-algebras on the unit interval coincide with complete MV-algebras, we have

Theorem 22 *Postulates* (V1) - (V8), (PI) *presuppose an injective MV-algebra structure on* $[0, 1]$*-valued Fuzzy Propositional Logic.*

From now on we assume, if not stated differently, that the set L of values of truth is a fixed injective MV-algebra. We set

Definition 22 *A function* $v : \mathcal{F} \searrow L$ *such that, for any inner truth value* $\mathbf{a}$ *and for any formulas* α, β,

$$v(\mathbf{a}) = a, \tag{3.1}$$
$$v(\alpha \texttt{ imp } \beta) = v(\alpha) \to v(\beta), \tag{3.2}$$
$$v(\alpha \texttt{ and } \beta) = v(\alpha) \odot v(\beta), \tag{3.3}$$

is called valuation *or* truth value function.

By the consideration above, any valuation v satisfies postulates (V1)-(V8), (PI). We introduce a logical connective or (read 'or' or 'disjunction') as an abbreviation

$$(\alpha \texttt{ or } \beta) = [\texttt{non-}(\texttt{non-}\alpha \texttt{ and } \texttt{non-}\beta)].$$

This generalizes the sate of affairs in Classical Propositional Logic and makes the formalized language of Fuzzy Propositional Logic easier to read.

Remark 20 *For any valuation v, any $\alpha, \beta \in \mathcal{F}$,*

$$v(\texttt{non-}\alpha) = v(\alpha)^*, \tag{3.4}$$

$$v(\alpha \texttt{ or } \beta) = v(\alpha) \oplus v(\beta). \tag{3.5}$$

Proof. Exercise. Generally, $v(\alpha \texttt{ and } \beta) < v(\alpha) \wedge v(\beta)$ and $v(\alpha) \vee v(\beta) < v(\alpha \texttt{ or } \beta)$. In application we may need, however, disjunctive and conjunctive connectives, denote them by $\overline{\texttt{and}}$ and $\overline{\texttt{or}}$, respectively, such that

$$v(\alpha \ \overline{\texttt{and}}\ \beta) = v(\alpha) \wedge v(\beta),\ v(\alpha \ \overline{\texttt{or}}\ \beta) = v(\alpha) \vee v(\beta).$$

Abbreviating can do this

$$(\alpha \ \overline{\texttt{or}}\ \beta) = [(\alpha \texttt{ imp } \beta) \texttt{ imp } \beta)]$$
$$(\alpha \ \overline{\texttt{and}}\ \beta) = [\texttt{non-}(\texttt{non-}\alpha \ \overline{\texttt{or}}\ \texttt{non-}\beta)],$$

even if the abbreviation of the logical connective $\overline{\texttt{or}}$ is far from being obvious. Indeed, we then have

$$v(\alpha \ \overline{\texttt{or}}\ \beta) = [v(\alpha) \to v(\beta)] \to v(\beta) = v(\alpha) \vee v(\beta),$$
$$v(\alpha \ \overline{\texttt{and}}\ \beta) = [v(\alpha)^* \vee v(\beta)^*]^* = v(\alpha) \wedge v(\beta).$$

We also introduce a logical connective equiv by abbreviating

$$(\alpha \texttt{ equiv } \beta) = [(\alpha \texttt{ imp } \beta) \ \overline{\texttt{and}}\ [(\beta \texttt{ imp } \alpha)],$$

thus generalizing the situation in Classical Propositional Logic. Then we have, for any valuation v, any formulas $\alpha, \beta \in \mathcal{F}$,

$$v(\alpha \texttt{ equiv } \beta) = v(\alpha) \leftrightarrow v(\beta).$$

In Classical Propositional Logic truth value functions $v : \mathcal{F} \searrow \{0, 1\}$ satisfy the truth tables

$v(\alpha \texttt{ or } \beta)$	$v(\beta) = 1$	$v(\beta) = 0$
$v(\alpha) = 1$	1	1
$v(\alpha) = 0$	1	0

$v(\alpha \texttt{ and } \beta)$	$v(\beta) = 1$	$v(\beta) = 0$
$v(\alpha) = 1$	1	0
$v(\alpha) = 0$	0	0

$v(\alpha$ imp $\beta)$	$v(\beta)=1$	$v(\beta)=0$
$v(\alpha)=1$	1	0
$v(\alpha)=0$	1	1

	$v(\alpha)=1$	$v(\alpha)=0$
$v($non-$\alpha)$	0	1

It is easy to verify that fuzzy valuations satisfy these tables and that in two valued case the truth value tables of the logical connectives $\overline{\text{and}}$ and and as well as $\overline{\text{or}}$ and or coincide. Complete truth tables are not, of course, possible in logic with infinite many values of truth. We may, however, write instances of them. For example, in the generalized Łukasiewicz structure, for $p = 3$, we have

$v(\beta)$

	$v(\alpha$ and $\beta)$	0.00	0.20	0.40	0.60	0.80	1.00
	0.00	0.00	0.00	0.00	0.00	0.00	0.00
	0.20	0.00	0.00	0.00	0.00	0.00	0.20
$v(\alpha)$	0.40	0.00	0.00	0.00	0.00	0.00	0.40
	0.60	0.00	0.00	0.00	0.00	0.00	0.60
	0.80	0.00	0.00	0.00	0.00	0.29	0.80
	1.00	0.00	0.20	0.40	0.60	0.80	1.00

$$v(\alpha \text{ and } \beta) = \sqrt[3]{\max\{0, v(\alpha)^3 + v(\beta)^3 - 1\}}$$

$v(\beta)$

	$v(\alpha$ or $\beta)$	0.00	0.20	0.40	0.60	0.80	1.00
	0.00	0.00	0.20	0.40	0.60	0.80	1.00
	0.20	0.20	0.25	0.42	0.61	0.80	1.00
$v(\alpha)$	0.40	0.40	0.42	0.50	0.65	0.83	1.00
	0.60	0.60	0.61	0.65	0.76	0.90	1.00
	0.80	0.80	0.80	0.83	0.90	1.00	1.00
	1.00	1.00	1.00	1.00	1.00	1.00	1.00

$$v(\alpha \text{ or } \beta) = \min\{1, \sqrt[3]{v(\alpha)^3 + v(\beta)^3}\}$$

$v(\beta)$

	$v(\alpha$ imp $\beta)$	0.00	0.20	0.40	0.60	0.80	1.00
	0.00	1.00	1.00	1.00	1.00	1.00	1.00
	0.20	0.99	1.00	1.00	1.00	1.00	1.00
$v(\alpha)$	0.40	0.98	0.98	1.00	1.00	1.00	1.00
	0.60	0.92	0.93	0.95	1.00	1.00	1.00
	0.80	0.79	0.79	0.82	0.89	1.00	1.00
	1.00	0.00	0.20	0.40	0.60	0.80	1.00

$$v(\alpha \text{ imp } \beta) = \min\{1, \sqrt[3]{1 - v(\alpha)^3 + v(\beta)^3}\}$$

Notice that here the first column is the truth table of non-α. We can also easily calculate that if $v(\alpha) = 0.2$, $v(\beta) = 0.6$ and $v(\gamma) = 0.9$ then, for example

$$v([\alpha \texttt{ imp } (\texttt{non-}\beta \texttt{or non-}\gamma)] \texttt{ imp } \gamma) = 0.9$$

In logic we are interested in the logical consequences of given statements. From semantic point of view this raises a question

Associating fixed values of truth to a set of well-formed formulas $T \subseteq \mathcal{F}$, what is the least degree of truth, or greatest lower bound of such degrees, of an arbitrary formula $\alpha \in \mathcal{F}$ with respect to T?

This leads us to the following semantic definitions

Definition 23 *A* fuzzy set T of formulas *is a function* $T : \mathcal{F} \searrow L$. *A truth value function* $v : \mathcal{F} \searrow L$ satisfies T *if* $T(\alpha) \leq v(\alpha)$ *for any formula* $\alpha \in \mathcal{F}$. *If there exists a valuation v such that v satisfies T the T is called* satisfiable.

The void set $\emptyset$ can be regarded as a fuzzy set of formulas by defining $\emptyset(\alpha) = 0$ for all formulas $\alpha \in \mathcal{F}$. The void set is of course satisfiable.

Definition 24 *The* degree of validity *of a formula $\alpha \in \mathcal{F}$ (with respect to a fuzzy set of formulas T) is a value*

$$\mathcal{C}^{\mathtt{sem}}(T)(\alpha) = \bigwedge\{v(\alpha) \mid v \textit{ satisfies } T\}. \tag{3.6}$$

In particular, if T is the void set we define the degree of tautology *of a formula α by*

$$\mathcal{C}^{\mathtt{sem}}(\alpha) = \bigwedge\{v(\alpha) \mid v \textit{ is a valuation }\}. \tag{3.7}$$

This definition is very natural and generalizes the concept of tautology in Classical Propositional Logic. Using g.l.b in definitions (3.6) and (3.7) explains partly why we have to assume the truth value set L forms a complete lattice. If $\mathcal{C}^{\mathtt{sem}}(T)(\alpha) = a$ we write $T \models_a \alpha$, in particular $\models_a \alpha$ if T is the void set. Of special interest will be formulas α such that $\models_1 \alpha$. Evidently, if $\models_1 \alpha$, then $T \models_1 \alpha$ for any fuzzy set T of formulas (even for those T which are not satisfied by any valuation v!)

When counting the degree of validity of a formula α, we are minimizing $v(\alpha)$ for such valuations v that satisfy T. How can this be done? We have already seen that formulas containing logical connectives non, or, $\overline{\texttt{and}}$, $\overline{\texttt{or}}$ and equiv are abbreviations of formulas containing only the logical connectives imp and **and** and the inner truth value **0**. Moreover, for any valuation v, any well-formed formulas α, β,

$$\begin{aligned} v(\texttt{non-}(\alpha \texttt{ imp non-}\beta)) &= [v(\alpha) \to v(\beta)^*]^* \\ &= [v(\alpha)^* \oplus v(\beta)^*]^* \\ &= v(\alpha) \odot v(\beta) \\ &= v(\alpha \texttt{ and } \beta), \end{aligned}$$

thus a formula (α **and** β) can be regarded as an abbreviation of a formula [non-(α imp non-β)]. We conclude that any well-formed formula α can be replaced by a formula β such that β contains no other connectives that imp and that $v(\alpha) = v(\beta)$ for any valuation v. Therefore, the problem of minimizing $v(\alpha)$ coincides with the problem of minimizing $v(\beta)$. The formation rules of well-formed formulas quarantee that in any such formula β there is, if at least one imp-connective, the unique *main* imp-connective, namely the last imp-connective which occurs when one constructs β from its parts. Thus, minimizing $v(\beta) = v(\beta_1 \texttt{ imp } \beta_2)$ leads to the problem of search for such a valuation v which maximizes $v(\beta_1)$ and minimizes $v(\beta_1)$, etc. For short formulas β this problem can be solved easily, while in general, we face here an open problem.

Example 1 Let the set of values of truth form the Łukasiewicz structure. We are interested in the degree of tautology of a formula α standing for

> *Assuming winter is long implies either winter is not cold or damages caused by floods are not serious, then insurance business is profitable.*

First we have to formalize α. Consider prositional variables p, q, r, s, where p corresponds to *winter is long*, q corresponds to *winter is cold* , r corresponds to *damages caused by floods are serious*, s corresponds to *insurance business is profitable.* Then α can be expressed by

$$[p \texttt{ imp } (\texttt{ non-}q \ \overline{\texttt{or}} \ \texttt{non-}r)] \texttt{ imp } s.$$

Now we should minimize $[v(p) \to (v(q)^* \vee v(r)^*)] \to v(s)$. Take $v(p) = v(q) = v(r) = v(r) = 0$. Then $v(\alpha) = [0 \to (0^* \vee 0^*)] \to 0 = 0$. Thus, $\models_0 \alpha$, i.e. the statement α has no general validity. Next consider a fuzzy set T of formulas such that $T(p) = 0.2$ (winter is practically not long at all), $T(q) = 0.8$ (winter is relatively cold), $T(r) = 0.1$ (damages caused by floods are minimal) $T(s) = 0.8$ (insurance business is quite profitable), $T(\beta) = 0$ elsewhere. T is satisfiable; indeed, a valuation v such that $v(p) = 0.2$, $v(q) = 0.8$, $v(r) = 0.1$ $v(s) = 0.8$ satisfies T. Moreover, $v(\alpha) = [0.2 \to (0.2 \vee 0.9)] \to 0.8 = 0.8$. This implies $C^{\mathtt{sem}}(T)(\alpha) \le 0.8$. On the other hand, since $v(s) \le v(\beta) \to v(s) = v(\beta \texttt{ imp } s) = v(\alpha)$, we reason that $0.8 \le C^{\mathtt{sem}}(T)(\alpha)$. Thus, $T \models_{0.8} \alpha$.

Roughly speaking, the meaning of a value $C^{\mathtt{sem}}(T)(\alpha)$ can be illustrated by saying

In all those worlds where any formula β is at least $T(\beta)$-true, a fixed formula α is $C^{\mathbf{sem}}(T)(\alpha)$-true.

A *consequence operation* $\mathcal{C}$ in Classical Propositional Logic, due to Tarski, is an operation $\mathcal{C} : \mathcal{F} \searrow \mathcal{F}$ such that, for any sets $X, Y \subseteq \mathcal{F}$, (i) $X \subseteq \mathcal{C}(X)$, (ii) if $X \subseteq Y$, then $\mathcal{C}(X) \subseteq \mathcal{C}(Y)$, (iii) $\mathcal{C}(X) = \mathcal{C}(\mathcal{C}(X))$. Since different two-valued logical systems such as Classical Propostional Logic, Intuitionistic Logic, Modal Logic, etc. can be characterized and distinguished by means of the sets of tautologies in them, logic is sometimes *defined* to be a consequence operation. The concept of consequence operation can be generalized into fuzzy logic in the following way; the collection $\tilde{\mathcal{F}}$ of all fuzzy sets of formulas is partially ordered by the relation $X \leq Y$ iff $X(\alpha) \leq Y(\alpha)$ for any $\alpha \in \mathcal{F}$. We set

Definition 25 *An operation $\mathcal{C} : \tilde{\mathcal{F}} \searrow \tilde{\mathcal{F}}$ is a* fuzzy consequence operation *if, for any $X, Y \in \tilde{\mathcal{F}}$,*
(C1) $X \leq \mathcal{C}(X)$,
(C2) *if* $X \leq Y$ *then* $\mathcal{C}(X) \leq \mathcal{C}(Y)$,
(C3) $\mathcal{C}(X) = \mathcal{C}(\mathcal{C}(X))$.

Proposition 86 *The operation $C^{\mathbf{sem}}$ is a fuzzy consequence operation, called* semantic consequence operation.

Proof. Let X, Y be fuzzy set of formulas. First we realize that any valuation v can be regarded as an element of $\tilde{\mathcal{F}}$ thus, by (3.6), $X \leq C^{\mathbf{sem}}(X)$. Hence (C1) holds. Clearly, if a valuation v satisfies X and $X \leq Y$ then v satisfies Y, too, thus (C2). Finally, $X \leq v$ iff $C^{\mathbf{sem}}(X) \leq v$, hence $\mathcal{C}(X) = \mathcal{C}(\mathcal{C}(X))$. □ Let T be a fuzzy set of formulas. A *finite part* of T is a fuzzy set of formulas S such that $T(\alpha) = S(\alpha)$ for finite many formulas $\alpha \in \mathcal{F}$ and $S(\alpha) = \mathbf{0}$ elsewhere. Clearly, any valuation v which satisfies T satisfies also S, thus, for any $\alpha \in \mathcal{F}$, $\bigwedge\{v(\alpha) \mid \ v \textit{ satisfies } S\} \leq \bigwedge\{v(\alpha) \mid \ v \textit{ satisfies } T\}$, i.e. $C^{\mathbf{sem}}(S) \leq C^{\mathbf{sem}}(T)$ for any finite part S of T. This implies

$$\bigvee\{C^{\mathbf{sem}}(S) \mid \ S \textit{ is a finite part of } T\} \leq C^{\mathbf{sem}}(T).$$

Definition 26 *A fuzzy consequence operation $\mathcal{C}$ is* continuous *if, for any fuzzy set of formulas T,*

$$\bigvee\{\mathcal{C}(S) \mid \ S \textit{ is a finite part of } T\} = \mathcal{C}(T) \tag{3.8}$$

Proposition 87 *Consider* $[0,1]$*-valued Fuzzy Propositional Logic such that the set of values of truth forms a residuated lattice but does not form an MV-algebra. Then the semantic consequence operation $C^{\mathbf{sem}}$ is not continuous.*

Proof. If $[0,1]$-valued residuated lattice is not an MV-algebra then either (2.93) or (2.94) does not hold. Assume (2.93) does not hold. By (1.67), there is an element $a \in [0,1]$ and an infinite subset $\{b_i\}_{i \in \Gamma} \subseteq [0,1]$ such that $\bigvee_{i\in\Gamma}(a \to b_i) = c < a \to (\bigvee_{i\in\Gamma} b_i)$. Let p be a fixed propositional variable and T a fuzzy set of formulas such that $T(\mathbf{b}_i \texttt{ imp } p) = 1$ for any $i \in \Gamma$ and $T(\alpha) = 0$ elsewhere. A valuation v_1 such that $v_1(p) = \bigvee_{i\in\Gamma} b_i$ satisfies T and any other valuation v which also satisfies T has the property $v_1(p) \leq v(p)$. Thus

$$\begin{aligned}
\mathcal{C}^{\text{sem}}(T)(\mathbf{a} \texttt{ imp } p) &= \bigwedge\{v(\mathbf{a} \texttt{ imp } p) \mid v \textit{ satisfies } T\}\\
&= \bigwedge\{v(\mathbf{a}) \to v(p) \mid v \textit{ satisfies } T\}\\
&= \bigwedge\{a \to v(p) \mid v \textit{ satisfies } T\}\\
&= a \to \bigwedge\{v(p) \mid v \textit{ satisfies } T\}\\
&= a \to v_1(p)\\
&= a \to (\bigvee_{i\in\Gamma} b_i).
\end{aligned}$$

Let S be a finite part of T. Then $S(\mathbf{b}_i \texttt{ imp } p) = 1$ for only finite many inner truth values $\mathbf{b}_i$. Since the unit interval is totally ordered there is $j \in \Gamma$ such that $b_j = \max\{b_i \in [0,1] \mid S(\mathbf{b}_i \texttt{ imp } p) = 1\}$. A valuation v_2 such that $v_2(p) = b_j$ satisfies S and any other valuation v which also satisfies S obeys the property $v_2(p) \leq v(p)$. Therefore

$$\begin{aligned}
\mathcal{C}^{\text{sem}}(S)(\mathbf{a} \texttt{ imp } p) &= \bigwedge\{v(\mathbf{a} \texttt{ imp } p) \mid v \textit{ satisfies } S\}\\
&= a \to v_2(p)\\
&= a \to b_j\\
&\leq c.
\end{aligned}$$

So $\bigvee\{\mathcal{C}^{\text{sem}}(S)(\mathbf{a} \texttt{ imp } p) \mid S \textit{ is a finite part of } T\} \leq c$ and (3.8) does not hold. Then assume (2.94) does not hold. By (1.66), there is an element $a \in [0,1]$ and an infinite subset $\{b_i\}_{i\in\Gamma} \subseteq [0,1]$ such that $\bigvee_{i\in\Gamma}(b_i \to a) = c < (\bigwedge_{i\in\Gamma} b_i) \to a$. Define a fuzzy set of formulas T by $T(p \texttt{ imp } \mathbf{b}_i) = 1$ for any $i \in \Gamma$, $T(\alpha) = 0$ elsewhere. A valuation v_3 such that $v_3(p) = \bigwedge_{i\in\Gamma} b_i$ satisfies T and any other valuation v which also satisfies T has the property $v(p) \leq v_3(p)$. Therefore

$$\begin{aligned}
\mathcal{C}^{\text{sem}}(T)(p \texttt{ imp } \mathbf{a}) &= \bigwedge\{v(p \texttt{ imp } \mathbf{a}) \mid v \textit{ satisfies } T\}\\
&= \bigwedge\{v(p) \to a \mid v \textit{ satisfies } T\}\\
&= \bigvee\{v(p) \mid v \textit{ satisfies } T\} \to a\\
&= v_3(p) \to a\\
&= (\bigwedge_{i\in\Gamma} b_i) \to a.
\end{aligned}$$

Yet, for any finite part S of T, there is always an index $j \in \Gamma$ such that $b_j = \min\{b_i \in [0,1] \mid S(p \texttt{ imp } \mathbf{b}_i) = 1\}$. A valuation v_4 such that $v_4(p) = b_j$ satisfies S and any other valuation v which also satisfies S fulfils $v(p) \leq v_4(p)$. We conclude

$$\begin{aligned}
\mathcal{C}^{\text{sem}}(S)(p \texttt{ imp } \mathbf{a}) &= \bigwedge\{v(p \texttt{ imp } \mathbf{a}) \mid v \textit{ satisfies } S\}\\
&= \bigwedge\{v(p) \to a \mid v \textit{ satisfies } S\}\\
&= \bigvee\{v(p) \mid v \textit{ satisfies } S\} \to a\\
&= v_4(p) \to a\\
&= b_j \to a\\
&\leq c.
\end{aligned}$$

Thus, $\bigvee\{C^{sem}(S)(p \texttt{ imp } \mathbf{a}) \mid \textit{ S is a finite part of T } \} \leq c$. Again, (3.8) does not hold. The proof is complete. □

Proposition 88 *Let α, β, γ, α_1, β_1, α_2, β_2 be formulas and* $\mathbf{c}$ *any inner truth value. Then the following forms of formulas are universally valid at the degree* $\mathbf{1}$*, except for, of course, the inner truth value* $\mathbf{c}$*, which is universally valid at the degree c, i.e.*

$$\models_1 \alpha \texttt{ imp } \alpha, \quad (3.9)$$

$$\models_1 (\alpha \texttt{ imp } \beta) \texttt{ imp } [(\beta \texttt{ imp } \gamma) \texttt{ imp } (\alpha \texttt{ imp } \gamma)], \quad (3.10)$$

$$\models_1 (\alpha_1 \texttt{imp} \beta_1) \texttt{ imp } \{(\beta_2 \texttt{imp} \alpha_2) \texttt{ imp } [(\beta_1 \texttt{imp} \beta_2) \texttt{ imp } (\alpha_1 \texttt{imp} \alpha_2)]\}, \quad (3.11)$$

$$\models_1 \alpha \texttt{ imp } \mathbf{1}, \quad (3.12)$$

$$\models_1 \mathbf{0} \texttt{ imp } \alpha, \quad (3.13)$$

$$\models_1 (\alpha \texttt{ and } \texttt{ non-}\alpha) \texttt{ imp } \mathbf{0}, \quad (3.14)$$

$$\models_c \mathbf{c}, \quad (3.15)$$

$$\models_1 \alpha \texttt{ imp } (\beta \texttt{ imp } \alpha), \quad (3.16)$$

$$\models_1 (\mathbf{1} \texttt{ imp } \alpha) \texttt{ imp } \alpha, \quad (3.17)$$

$$\models_1 [(\alpha \texttt{ imp } \beta) \texttt{ imp } \beta] \texttt{ imp } [(\beta \texttt{ imp } \alpha) \texttt{ imp } \alpha], \quad (3.18)$$

$$\models_1 (\texttt{non-}\alpha \texttt{ imp } \texttt{ non-}\beta) \texttt{ imp } (\beta \texttt{ imp } \alpha). \quad (3.19)$$

Proof. Exercise.

Remark 21 *Notice that Proposition 88 holds for a general residuated lattice.*

3.2 Axiom System for Fuzzy Logic

The definitions of valuation and the degree of validity of formula α are natural and relatively easy generalizations of the corresponding concepts in two valued logic. Now we consider the following related non-trivial problem

Knowing that a formula α is valid at a certain degree, do there exist a fuzzy set of axioms and fuzzy rules of inference by which we can infer α at the same degree?

In other words, is there a syntactic consequence operation that coincides with the semantic consequence operation, i.e. is Fuzzy Propositional Logic axiomatizable? To find an answer to this question, we start by defining on what we mean by fuzzy axiom, fuzzy rule of inference, fuzzy proof, etc.

A *rule of inference* in Classical Propositional Logic is an n-ary operation on the set of well-formed formulas which with a finite sequence of formulas $\alpha_1, \cdots, \alpha_n$ $(1 \leq n)$ in a formalized language associates another formula β in this language in such a way that β is a logical consequence of the formulas $\alpha_1, \cdots, \alpha_n$. This fact is usually denoted as follows

$$\frac{\alpha_1, \cdots, \alpha_n}{\beta}$$

Formulas $\alpha_1, \cdots, \alpha_1$ are called *premises* and β the *conclusion* of this rule of inference. For example,

$$\frac{\texttt{non} - (\texttt{non} - \alpha)}{\alpha}$$

and

$$\frac{\alpha, (\alpha \texttt{ imp } \beta)}{\beta}$$

are rules of inference in Classical Propositional Logic, called *Rule of Double Negation* and *Modus Ponens,* respectively. By saying that a formula β is a *logical consequence* of a set S of formulas we mean that if every formula α belonging to S is acknowledged to be true, then β must be accepted as true. Thus, the most important property of rule of inference is soundness, i.e. rule of inference preserves truth.

We define a fuzzy rule of inference as consisting of two components. The first component operates on formulas and is, in fact, a rule of inference in the usual sense; the second component operates on truth values and says how the truth value of the conclusion is to be computed from the truth values of the premises such that the degree of truth is preserved. More accurately, we set

Definition 27 *An n-ary* fuzzy rule of inference *is a scheme*

$$R \quad : \quad \frac{\alpha_1, \cdots, \alpha_n}{r^{\text{syn}}(\alpha_1, \cdots, \alpha_n)}, \quad \frac{a_1, \cdots, a_n}{r^{\text{sem}}(a_1, \cdots, a_n)}$$

where the well-formed formulae $\alpha_1, \cdots, \alpha_n$ are the premises *and the well-formed formula $r^{\text{syn}}(\alpha_1, \cdots, \alpha_n)$ is the* conclusion. *The values $a_1, \cdots, a_n$, $r^{\text{sem}}(a_1, \cdots, a_n) \in L$ are the corresponding truth values. The mapping $r^{\text{sem}} : L^n \searrow L$ is* semi-continuous *on each variable, i.e. it holds always that*

$$r^{\text{sem}}(a_1, \cdots, \bigvee_{j \in \Gamma} a_{k_j}, \cdots, a_n) = \bigvee_{j \in \Gamma} r^{\text{sem}}(a_1, \cdots, a_{k_j}, \cdots, a_n), 1 \leq k \leq n.$$

We assume the fuzzy rule of inference is sound, *i.e. for each valuation v holds*

$$r^{\text{sem}}(v(\alpha_1), \cdots, v(\alpha_n)) \leq v(r^{\text{syn}}(\alpha_1, \cdots, \alpha_n)).$$

Remark 22 *r^{sem} is isotone in each variable.*

Proof. Indeed, let $a \leq b$. Then $a \vee b = b$ and

$$
\begin{aligned}
r^{\mathtt{sem}}(a_1,\cdots,a,\cdots,a_n) &\leq r^{\mathtt{sem}}(\cdots,a,\cdots)\vee r^{\mathtt{sem}}(\cdots,b,\cdots)\\
&= r^{\mathtt{sem}}(\cdots,a\vee b,\cdots)\\
&= r^{\mathtt{sem}}(a_1,\cdots,b,\cdots,a_n).\square
\end{aligned}
$$

The opposite of Remark 22 is, in general, not true as we shall see in Exercise 1.

Proposition 89 *Consider complete MV-algebra valued Fuzzy Propositional Logic. The following schemes are fuzzy rules of inference*

Generalized Modus Ponens:

$$R_{GMP} \quad : \quad \frac{\alpha, (\alpha \ \mathtt{imp}\ \beta)}{\beta}, \ \frac{a, b}{a \odot b}$$

a-Consistency-testing rules:

$$R_{a-CTR} \quad : \quad \frac{\mathbf{a}}{\mathbf{0}}, \ \frac{b}{c}$$

where **a** *is an inner truth value, and* $c = \mathbf{0}$ *if* $b \leq a$ *and* $c = \mathbf{1}$ *elsewhere.*
a-Lifting Rules:

$$R_{a-LR} \quad : \quad \frac{\alpha}{(\mathbf{a}\ \mathtt{imp}\ \alpha)}, \ \frac{b}{a \to b}$$

where **a** *is an inner truth value.*
Rule of Bold Conjunction:

$$R_{RBC} \quad : \quad \frac{\alpha, \beta}{(\alpha \ \mathbf{and}\ \beta)}, \ \frac{a, b}{a \odot b}$$

Proof. First we verify soundness. Let v *be a valuation. Then*

$$
\begin{aligned}
r^{\mathtt{sem}}(v(\alpha), v(\alpha\ \mathtt{imp}\ \beta)) &= v(\alpha)\odot v(\alpha\ \mathtt{imp}\ \beta)\\
&= v(\alpha)\odot [v(\alpha)\to v(\beta)]\\
&\leq v(\beta)\\
&= v(r^{\mathtt{syn}}(\alpha, [\alpha\ \mathtt{imp}\ \beta])).
\end{aligned}
$$

Hence, Generalized Modus Ponens is sound. Then we reason

$$
\begin{aligned}
r^{\mathtt{sem}}(v(\mathbf{a})) &= r^{\mathtt{sem}}(a)\\
&= \mathbf{0}\\
&= v(\mathbf{0})\\
&\leq v(r^{\mathtt{syn}}(a)).
\end{aligned}
$$

implying **a**-Consistency-testing rules are sound. For **a**-Lifting Rules we have

$$
\begin{aligned}
r^{\mathtt{sem}}(v(\alpha)) &= \mathbf{a}\to v(\alpha)\\
&= v(\mathbf{a}\ \mathtt{imp}\ \alpha)\\
&\leq v(r^{\mathtt{syn}}(\alpha)).
\end{aligned}
$$

Finally, for Rule of Bold Conjunction soundness holds by the fact

$$\begin{aligned} r^{\mathtt{sem}}(v(\alpha), v(\beta)) &= v(\alpha) \odot v(\beta) \\ &= v(\alpha \text{ and } \beta) \\ &\leq v(r^{\mathtt{syn}}(\alpha, \beta)). \end{aligned}$$

To prove that these schemas are semi-continuous we reason
-for Generalized Modus Ponens and Rule of Bold Conjunction,

$$r^{\mathtt{sem}}(a, \bigvee_{i\in\Gamma} b_i) = a \odot (\bigvee_{i\in\Gamma} b_i) = \bigvee_{i\in\Gamma}(a \odot b_i) = \bigvee_{i\in\Gamma} r^{\mathtt{sem}}(a, b_i),$$

-for **a**-Consistency-testing rules,
either $b_i \not\leq a$ for some $i \in \Gamma$ whence $\bigvee_{i\in\Gamma} b_i \not\leq a$ implying

$$r^{\mathtt{sem}}(\bigvee_{i\in\Gamma} b_i) = \mathbf{1} = \bigvee_{i\in\Gamma} r^{\mathtt{sem}}(b_i)$$

or $b_i \leq a$ for all $i \in \Gamma$ whence $\bigvee_{i\in\Gamma} b_i \leq a$ implying

$$r^{\mathtt{sem}}(\bigvee_{i\in\Gamma} b_i) = \mathbf{0} = \bigvee_{i\in\Gamma} r^{\mathtt{sem}}(b_i),$$

-for **a**-Lifting Rules,

$$r^{\mathtt{sem}}(\bigvee_{i\in\Gamma} b_i) = \mathbf{a} \to \bigvee_{i\in\Gamma} b_i = \bigvee_{i\in\Gamma}(\mathbf{a} \to b_i) = \bigvee_{i\in\Gamma} r^{\mathtt{sem}}(b_i).$$

The proof is complete. □

Notice that, in general residuated lattice valued Fuzzy Propositional Logic, the **a**-Lifting Rules are not fuzzy rules of inference as the semi-continuous condition cease to hold there.

Definition 28 *A fuzzy set* $\mathtt{A}$ *of* logical axioms *in Fuzzy Propositional Logic is a finite set of forms of formulas each being an inner truth value* **a**, *then* $\mathtt{A}(\mathbf{a}) = a$, *or a tautology* α *at the degree* $\mathbf{1}$, *then* $\mathtt{A}(\alpha) = \mathbf{1}$. *Elsewhere* $\mathtt{A}(\alpha) = \mathbf{0}$.

For example, the following set of forms of formulas is, by (3.9)-(3.19), a set of logical axioms:
(Ax.1) α imp α,
(Ax.2) $(\alpha$ imp $\beta)$ imp $[(\beta$ imp $\gamma)$ imp $(\alpha$ imp $\gamma)]$,
(Ax.3) $(\alpha_1$ imp $\beta_1)$ imp $\{(\beta_2$ imp $\alpha_2)$ imp $[(\beta_1$ imp $\beta_2)$ imp $(\alpha_1$ imp $\alpha_2)]\}$,
(Ax.4) α imp $\mathbf{1}$,
(Ax.5) $\mathbf{0}$ imp α,
(Ax.6) $[(\alpha$ and non-$\alpha)$ imp $\mathbf{0}$,
(Ax.7) **a**,
(Ax.8) α imp $(\beta$ imp $\alpha)$,
(Ax.9) $(\mathbf{1}$ imp $\alpha)$ imp α,
(Ax.10) $[(\alpha$ imp $\beta)$ imp $\beta]$ imp $[(\beta$ imp $\alpha)$ imp $\alpha]$,
(Ax.11) (non-α imp non-$\beta)$ imp $(\beta$ imp $\alpha)$.
where α, β, γ, α_1, α_2, β_1, β_2 are well-formed formulas and **a** any inner truth value. Values $\mathtt{A}(\delta)$ are obvious.

Definition 29 *Let* A *be a fixed set of logical axioms,* R *a fixed finite set of fuzzy rules of inference and* T *fuzzy set of formulas called* non-logical axioms. *Then a (zero-order)* fuzzy theory *is a triplet* $\langle \mathtt{A}, \mathtt{R}, T\rangle$. *In particular, if the set of logical axioms* A *is composed of Ax.1 - Ax.11, and the set of fuzzy rules of inference* R *contains* R_{GMP}, R_{a-CTR}, R_{a-LR}, R_{RBC}, *we denote a fuzzy theory simply by* T, *and if* T *is the void set we talk about* Fuzzy Propositional Calculus.

A *metaproof* of a well-formed formula α in a fuzzy theory $\langle \mathtt{A}, \mathtt{R}, T\rangle$, denoted by w, is a finite sequence

$$\begin{array}{ccc} \alpha_1 & , & a_1 \\ \vdots & & \vdots \\ \alpha_m & , & a_m \end{array}$$

of pairs $(\alpha_i, a_i) \in \mathcal{F} \times L$ such that the following holds: (i) $\alpha_m = \alpha$, (ii) for each $i, 1 \leq i \leq m, \alpha_i$ is a logical axiom, or α_i is a non-logical axiom, or there are a fuzzy rule of inference in R and formulas $\alpha_{i_1}, \cdots, \alpha_{i_n}$ with $i_1, \cdots, i_n < i$ such that $\alpha_i = r^{\mathtt{syn}}(\alpha_{i_1}, \cdots, \alpha_{i_n})$, (iii) for each $i, 1 \leq i \leq m$, the value a_i is given by

$$a_i = \begin{cases} a & \text{if } \alpha_i \text{ is the axiom } \mathbf{a} \\ \mathbf{1} & \text{if } \alpha_i \text{ is some other logical axiom} \\ T(\alpha_i) & \text{if } \alpha_i \text{ is a non-logical axiom} \\ r^{\mathtt{sem}}(a_{i_1}, \cdots, a_{i_n}) & \text{if } \alpha_i = r^{\mathtt{syn}}(\alpha_{i_1}, \cdots, \alpha_{i_n}). \end{cases}$$

The value a_m is denoted by $Val_{\langle \mathtt{A},\mathtt{R},T\rangle}(w)$ and is called the *degree* of the metaproof w. Because a formula α may have many metaproofs with different degrees, we define the *degree of deduction* of the formula α in fuzzy theory $\langle \mathtt{A}, \mathtt{R}, T\rangle$ by

$$C^{\mathtt{syn}\langle \mathtt{A},\mathtt{R}\rangle}(T)(\alpha) = \bigvee\{ Val_{\langle \mathtt{A},\mathtt{R},T\rangle}(w) | w \textit{ is a metaproof for } \alpha \textit{ in } \langle \mathtt{A}, \mathtt{R}, T\rangle\}.$$

The case $C^{\mathtt{syn}\langle \mathtt{A},\mathtt{R}\rangle}(T)(\alpha) = a$ is denoted by $\langle \mathtt{A}, \mathtt{R}, T\rangle \vdash_a \alpha$, in particular, $\vdash_a \alpha$ if the set of logical axioms A is composed of Ax.1 - Ax.11, the set of fuzzy rules of inference R contains R_{GMP}, R_{a-CTR}, R_{a-LR}, R_{RBC} and T is the void set.

The following two propositions hold for any residuated lattice valued Fuzzy Propositional Logic.

Proposition 90 *The operation* $C^{\mathtt{syn}\langle \mathtt{A},\mathtt{R}\rangle}$ *is a fuzzy consequence operation, called* syntactic consequence operation.

Proof. Clearly, $C^{\mathtt{syn}\langle \mathtt{A},\mathtt{R}\rangle} : \tilde{\mathcal{F}} \searrow \tilde{\mathcal{F}}$. Let X, Y be fuzzy sets of formulas such that $X \leq Y$. Since, for any $\alpha \in \mathcal{F}$,

$$\alpha \quad , \quad X(\alpha) \leq Y(\alpha)$$

can be regarded as a metaproof for α

in the fuzzy theory $\langle \mathtt{A}, \mathtt{R}, X\rangle$ or $\langle \mathtt{A}, \mathtt{R}, Y\rangle$, respectively, we have $X(\alpha) \leq C^{\mathtt{syn}\langle \mathtt{A},\mathtt{R}\rangle}(X)(\alpha) \leq C^{\mathtt{syn}\langle \mathtt{A},\mathtt{R}\rangle}(Y)(\alpha)$. By a similar argument, $\mathtt{A}(\alpha) \leq C^{\mathtt{syn}\langle \mathtt{A},\mathtt{R}\rangle}(X)(\alpha)$. In particular,

$$C^{\mathtt{syn}\langle \mathtt{A},\mathtt{R}\rangle}(X) \leq C^{\mathtt{syn}\langle \mathtt{A},\mathtt{R}\rangle}(C^{\mathtt{syn}\langle \mathtt{A},\mathtt{R}\rangle}(X)).$$

Let $\alpha \in \mathcal{F}$. Any metaproof w for α in the fuzzy theory $\langle \mathtt{A}, \mathtt{R}, C^{\mathtt{syn}\langle \mathtt{A},\mathtt{R}\rangle}(X)\rangle$ is of form

$$\begin{array}{ccc} \alpha_1 & , & a_1 \\ \vdots & & \vdots \\ \alpha_m & , & a_m \end{array}$$

For each $i \leq m$, if α_i is a logical or non-logical axiom, then obviously $a_i \leq C^{\mathtt{syn}\langle \mathtt{A},\mathtt{R}\rangle}(X)(\alpha_i)$, and if $\alpha_i = r^{\mathtt{syn}}(\alpha_{i_1}, \cdots, \alpha_{i_n})$, then $a_i = r^{\mathtt{sem}}(a_{i_1}, \cdots, a_{i_n})$, where, for $1 \leq j \leq n$,

$$a_{i_j} \leq \bigvee\{Val_{\langle \mathtt{A},\mathtt{R},X\rangle}(w_{i_j}) \mid w_{i_j} \text{ is a metaproof for } \alpha_{i_j} \text{ in } \langle \mathtt{A}, \mathtt{R}, X\rangle\}.$$

By semi-continuity,

$$\begin{array}{rcl} a_i & \leq & \bigvee\{r^{\mathtt{sem}}(Val_{\langle \mathtt{A},\mathtt{R},X\rangle}(w_{i_1}), \cdots, Val_{\langle \mathtt{A},\mathtt{R},X\rangle}(w_{i_n}))\} \\ & = & C^{\mathtt{syn}\langle \mathtt{A},\mathtt{R}\rangle}(X)(\alpha_i). \end{array}$$

In particular, $Val_{\langle \mathtt{A},\mathtt{R},C^{\mathtt{syn}\langle \mathtt{A},\mathtt{R}\rangle}\rangle}(w) \leq C^{\mathtt{syn}\langle \mathtt{A},\mathtt{R}\rangle}(X)(\alpha)$. Therefore

$$\bigvee\{Val_{\langle \mathtt{A},\mathtt{R},C^{\mathtt{syn}\langle \mathtt{A},\mathtt{R}\rangle}\rangle}(w) \mid w \text{ is a metaproof for } \alpha \text{ in } \langle \mathtt{A}, \mathtt{R}, C^{\mathtt{syn}\langle \mathtt{A},\mathtt{R}\rangle}(X)\rangle\} \leq C^{\mathtt{syn}\langle \mathtt{A},\mathtt{R}\rangle}(X)(\alpha).$$

This proves $C^{\mathtt{syn}\langle \mathtt{A},\mathtt{R}\rangle}(C^{\mathtt{syn}\langle \mathtt{A},\mathtt{R}\rangle}(X)) \leq C^{\mathtt{syn}\langle \mathtt{A},\mathtt{R}\rangle}(X)$. Thus, $C^{\mathtt{syn}\langle \mathtt{A},\mathtt{R}\rangle}(X) = C^{\mathtt{syn}\langle \mathtt{A},\mathtt{R}\rangle}(C^{\mathtt{syn}\langle \mathtt{A},\mathtt{R}\rangle}(X))$. The proof is complete. □

Proposition 91 *The consequence operation* $C^{\mathtt{syn}\langle \mathtt{A},\mathtt{R}\rangle}$ *is a continuous.*

Proof. Obviously, if S is a finite part of T, then

$$C^{\mathtt{syn}\langle \mathtt{A},\mathtt{R}\rangle}(S)(\alpha) \leq C^{\mathtt{syn}\langle \mathtt{A},\mathtt{R}\rangle}(T)(\alpha).$$

Thus,

$$\bigvee\{C^{\mathtt{syn}\langle \mathtt{A},\mathtt{R}\rangle}(S) \mid \textit{S is a finite part of T}\} \leq C^{\mathtt{syn}\langle \mathtt{A},\mathtt{R}\rangle}(T).$$

Let $\alpha \in \mathcal{F}$ and let w be a metaproofs for α in the fuzzy theory $\langle \mathtt{A}, \mathtt{R}, T\rangle$. Since w is finite we may construct a finite part S_w of T by setting $S_w(\alpha_k) = T(\alpha_k)$ if in w occurs

$$\alpha_k \quad , \quad T(\alpha_k)$$

and $S_w(\beta) = \mathbf{0}$ elsewhere. Then $Val_{\langle \mathtt{A},\mathtt{R},T\rangle}(w) \leq C^{\mathtt{syn}\langle \mathtt{A},\mathtt{R}\rangle}(S_w)(\alpha)$. Therefore

$$\begin{aligned} C^{syn\langle A,R\rangle}(T)(\alpha) &\leq \bigvee\{C^{syn\langle A,R\rangle}(S_w)(\alpha) \mid w \textit{ metaproof for } \alpha \text{ in } \langle A, R, S_w\rangle\} \\ &\leq \bigvee\{C^{syn\langle A,R\rangle}(S)(\alpha) \mid S \textit{ is a finite part of } T\} \\ &\leq C^{syn\langle A,R\rangle}(T)(\alpha). \end{aligned}$$

We conclude $C^{syn\langle A,R\rangle}(T) = \bigvee\{C^{syn\langle A,R\rangle}(S) \mid S \textit{ is a finite part of } T\}$. The proof is complete. □

Fix the truth value set L of Fuzzy Propostional Logic. If $C^{syn\langle A,R\rangle}$ is continuous while C^{sem} is not then, of course, the corresponding fuzzy theories are not complete. Now, necessary condition for continuity of semantic consequence operation C^{sem} in $[0,1]$-valued Fuzzy Propositional Logic is, by Proposition 87, that the truth value set $[0,1]$ forms a complete MV-algebra. Thus, we have

Theorem 23 (Incompleteness Theorem for Fuzzy Logic). *Assume the truth value set in* $[0,1]$*-valued Fuzzy Propositional Logic forms a residuated lattice but does not form an complete MV-algebra. Then there exist fuzzy theories T which are not axiomatizable, i.e.* $C^{sem}(T) \neq C^{syn}(T)$.

For now on, we assume that the set L of values of truth is an injective MV-algebra. Moreover, the set of logical axioms A is composed of Ax.1 - Ax.11, and the set of fuzzy rules of inference R contains the fuzzy rules of inference R_{GMP}, R_{a-CTR}, R_{a-LR}, R_{RBC}. Fuzzy theories are thus identified by means of their sets T of non-logical axioms. We will write C^{syn} instead of $C^{syn\langle A,R\rangle}$. Obviously, for any fuzzy theory T, if $\vdash_1 \alpha$ then $T \vdash_1 \alpha$, and by Ax.7, for any inner truth value $\mathbf{a}$, $a \leq C^{syn}(T)(\mathbf{a})$. This leads us to the following

Definition 30 *A fuzzy theory T is* consistent *if, for any inner truth value* $\mathbf{a}$, $a = C^{syn}(T)(\mathbf{a})$, *and otherwise T is* contradictory.

Proposition 92 *A fuzzy theory T is contradictory iff $T \vdash_1 \alpha$ holds for each $\alpha \in \mathcal{F}$.*

Proof. Assume T is contradictory. Then there exists an inner truth value $\mathbf{a}$ such that $a \neq C^{syn}(T)(\mathbf{a})$. If for each metaproof w for $\mathbf{a}$ holds $Val_T(w) \leq a$, then $a \leq C^{syn}(T)(\mathbf{a}) \leq a$, hence $C^{syn}(T)(\mathbf{a}) = a$, which is not the case. Therefore there exists a metaproof w for $\mathbf{a}$ such that $Val_T(w) \not\leq a$. For every formula $\alpha \in \mathcal{F}$, we have now the following metaproof:

$\mathbf{a}$,	$Val_T(w)$,	*assumption*
$\mathbf{0}$,	1 ,	R_{a-CTR}
$\mathbf{0}$ imp α ,	1 ,	*Ax*.4
α ,	1 ,	R_{GMP}

We conclude that $T \vdash_1 \alpha$ holds for each $\alpha \in \mathcal{F}$. Conversely, if $T \vdash_1 \alpha$ holds for each $\alpha \in \mathcal{F}$, then, in particular, $T \vdash_1 \mathbf{0}$, i.e. $C^{syn}(T)(\mathbf{0}) = 1 \neq \mathbf{0}$. □

Proposition 93 *A fuzzy theory T is contradictory iff the following condition holds* (C)*:*

There is a formula α and metaproofs w, w' for α, non-α*, respectively, such that $Val_T(w) = a$, $Val_T(w') = b$ and $\mathbf{0} < a \odot b$.*

Proof. Assume (C) holds. Let β be an arbitrary formula. There is the following metaproof for β in T:

α	, a	, *assumption*
non-α	, b	, *assumption*
(non-α and α)	, $a \odot b$	, R_{RBC}
(non-α and α) imp $\mathbf{0}$	, $\mathbf{1}$	, $Ax.6$
$\mathbf{0}$	, $a \odot b$	, R_{GMP}
$\mathbf{0}$	, $\mathbf{1}$	, $R_{\mathbf{0}-CTR}$
$\mathbf{0}$ imp β	, $\mathbf{1}$	, $Ax.4$
β	, $\mathbf{1}$	, R_{GMP}

We conclude that $T \vdash_1 \beta$ holds for each $\beta \in \mathcal{F}$, thus T is contradictory. Conversely, assume T is contradictory. Then, for each formula α, $T \vdash_1 \alpha$ and $T \vdash_1$ non-α. Consider sets

$$A = \{a \mid Val_T(w) = a \neq \mathbf{0}, w \textit{ is a metaproof for } \alpha\},$$
$$B = \{b \mid Val_T(w) = b \neq \mathbf{0}, w \textit{ is a metaproof for } \text{non-}\alpha\}.$$

Assume $a \odot b = \mathbf{0}$ for all $a \in A$, $b \in B$. Then $\mathbf{0} = \bigvee\{a \odot b \mid b \in B\} = a \odot \bigvee\{b \mid b \in B\} = a \odot \mathbf{1} = a$ for every $a \in A$, a contradiction. Thus, (C) holds. □

Let T be a fuzzy theory. The choice of the logical axioms Ax.1 - Ax.11 and soundness of fuzzy rules of inference guarantee, for each formula α, each metaproof w for α in T, each valuation v which satisfies T, that $Val_T(w) \leq v(\alpha)$. Thus,

$$\bigvee\{Val_T(w) \mid w \textit{ is a metaproof for } \alpha \textit{ in } T\} \leq \bigwedge\{v(\alpha) \mid v \textit{ satisfies } T\},$$

by symbols, $\mathcal{C}^{syn}(T)(\alpha) \leq \mathcal{C}^{sem}(T)(\alpha)$. (This (in-)equality holds even if T is not satisfiable as $\bigwedge\{\emptyset\} = \mathbf{1}$.) We write

Theorem 24 (Soundness Theorem for Fuzzy Propositional Logic) *Let T be a fuzzy theory. For each formula α, if $T \vdash_a \alpha$, $T \models_b \alpha$, then $a \leq b$.*

Proposition 94 *Any satisfiable fuzzy theory T is consistent.*

Proof. Let v satisfy T, $\alpha \in \mathcal{F}$ and $v(\alpha) = c$. Then $v(\text{non-}\alpha) = c^*$. Assume $T \vdash_a \alpha$, $T \vdash_b$ non-α. Then

$$a = \bigvee\{Val_T(w) \mid w \textit{ is a metaproof for } \alpha \textit{ in } T\} \leq v(\alpha) = c.$$

and

$$b = \bigvee \{Val_T(w) \mid w \textit{ is a metaproof for } \textbf{non-}\alpha \textit{ in } T\} \leq v(\textbf{non} - \alpha) = c^*.$$

Thus, for any metaproofs w, w' for α, $\textbf{non} - \alpha$, respectively, holds

$$Val_T(w) \odot Val_T(w') \leq a \odot b \leq c \odot c^* = \mathbf{0}.$$

We conclude that T is not contradictory and is therefore consistent. □

Corollary Fuzzy Propositional Calculus is consistent.

Proposition 95 *Let T be a fuzzy theory, α a well-formed formula and $a, b \in L$. If $T \vdash_1 (\mathbf{a}\ \texttt{imp}\ \alpha)$, $T \vdash_b \alpha$, then $a \leq b$.*

Proof. Let w be a metaproof for the formula $(\mathbf{a}\ \texttt{imp}\ \alpha)$ and $Val_T(w) = c$. We write the following metaproof w' for α:

$$\begin{array}{lcccc} \mathbf{a}\ \texttt{imp}\ \alpha & , & c & , & \textit{assumption} \\ \mathbf{a} & , & a & , & Ax.7 \\ \alpha & , & c \odot a & , & R_{GMP} \end{array}$$

Hence, we have

$$\begin{array}{rcl} a = a \odot 1 & = & a \odot \bigvee\{Val_T(w) \mid w \textit{ is a metaproof for } (\mathbf{a}\ \texttt{imp}\ \alpha)\} \\ & = & \bigvee\{a \odot Val_T(w) \mid w \textit{ is a metaproof for } (\mathbf{a}\ \texttt{imp}\ \alpha)\} \\ & \leq & \bigvee\{Val_T(w') \mid w' \textit{ is a metaproof for } \alpha\} \\ & = & b. \end{array}$$

□

Proposition 96 *Assume T be a fuzzy theory, α a well-formed formula, and $T \vdash_a \alpha$. Then $T \vdash_1 (\mathbf{a}\ \texttt{imp}\ \alpha)$.*

Proof. Let w be a metaproof for the formula α and $Val_T(w) = c$. We write the following metaproof w' for $(\mathbf{a}\ \texttt{imp}\ \alpha)$:

$$\begin{array}{lcccc} \alpha & , & c & , & \textit{assumption} \\ \mathbf{a}\ \texttt{imp}\ \alpha & , & a \to c & , & R_{\mathbf{a}-LR} \end{array}$$

Hence, we have

$$\begin{array}{rcl} 1 = a \to a & = & a \to \bigvee\{Val_T(w) \mid w \textit{ is a metaproof for } \alpha\} \\ & = & \bigvee\{a \to Val_T(w) \mid w \textit{ is a metaproof for } \alpha\} \\ & \leq & \bigvee\{Val_T(w') \mid w' \textit{ is a metaproof for } (\mathbf{a}\ \texttt{imp}\ \alpha)\}. \end{array}$$

We conclude $T \vdash_1 (\mathbf{a}\ \texttt{imp}\ \alpha)$. □

Finiteness Theorem in Classical Propositional Logic states that if a formula α is deducible from a set T of formulas, i.e. if there is a metaproof for α in T, then α is deducible from a finite part S of T. As one may assume, this is not, in general, the case in Fuzzy Propositional Logic as we shall now see.

Consider Łukasiewicz valued Fuzzy Propositional Logic. Let T be a fuzzy theory such that $T(\mathbf{a}_n \texttt{ imp } p) = 1$, where p is a fixed propositional variable, $a_n = \frac{n-1}{n}$, $n \in \mathcal{N}$, and $T(\alpha) = 0$ elsewhere. Obviously, any valuation v such that $v(p) = 1$ satisfies T, thus T is consistent. Let $0 \leq b < 1$ and $\frac{1}{1-b} < n$. Then $b < a_n$. We write the following metaproof w for p:

$$\begin{array}{llr} \mathbf{a}_n \texttt{ imp } p & , \mathbf{1} , & \textit{non-logical axiom} \\ \mathbf{a}_n & , a_n , & Ax.7 \\ p & , a_n , & R_{GMP} \end{array}$$

i.e. $Val_T(w) = a_n$. Hence, $\bigvee\{Val_T(w) \mid w \textit{ is a metaproof for } p \textit{ in } T\}$ $= 1$. Thus $T \vdash_1 p$. Assume now S is a finite part of T and

$$a_i = \max\{a_k \mid S(\mathbf{a}_k \texttt{ imp } p) = 1\} < 1.$$

Then a valuation v such that $v(p) = a_i$ satisfies S, thus S is consistent and $S \models_c p$, where $c \leq a_i$. By Soundness Theorem, $S \vdash_d p$, where $d \leq c < 1$. We conclude that $S \vdash_1 p$ does not hold for any finite part S of T.

We conclude this section by having a look at some application of Łukasiewicz valued Fuzzy Propositional Logic.

Example 1 Assume p stands for *It is raining enough* and q stands for *Potato is growing fast.* We study a fuzzy theory T such that $T(\texttt{non} - p \texttt{ imp non} - q) = 1$ standing for *If it is not raining enough then potato is not growing fast* and $T(q) = 0.7$ standing loosely for *Potato is growing more or less fast.* Now we are interested in the degree of deduction of p. We find the following metaproof for p:

$$\begin{array}{lccr} (\texttt{non} - p \texttt{ imp non} - q) \texttt{ imp } (q \texttt{ imp } p) & , & 1 , & Ax.11 \\ (\texttt{non} - p \texttt{ imp non} - q) & , & 1 , & \textit{non-logical axiom} \\ (q \texttt{ imp } p) & , & 1 , & R_{GMP} \\ q & , & 0.7 , & \textit{non-logical axiom} \\ p & , & 0.7 , & R_{GMP} \end{array}$$

Therefore $0.7 \leq \mathcal{C}^{\texttt{syn}}(p)$. Since a valuation v such that $v(p) = v(q) = 0.7$ satisfies T, we have $0.7 \leq \mathcal{C}^{\texttt{syn}}(T)(p) \leq \mathcal{C}^{\texttt{sem}}(T)(p) \leq 0.7$. Thus, the degree of deduction of p is 0.7. Freely speaking, *Is is raining more or less enough.*

Example 2 Assume Bob would like to get to know Mary. To do this, he should make a good impression on her. Unfortunately, Bob is shy with girls when being sober, thus, makes no impression on Mary. When Bob is drunk the impression he makes is not good. Let p stand for *Bob is sober,* q stand for *Bob makes a good impression on Mary,* and r stand for *Bob gets to know Mary.* Then the above state of affairs can be expressed formally in the following way: (A) (p `imp non-`q); When sober Bob does not make a good impression on Mary,

(B) (non-p imp non-q); When drunk Bob does not make a good impression on Mary,

(C) [(p $\overline{\text{and}}$ non-p) imp q]; Only drunk and sober Bob can make a good impression on Mary,

(D) [(q or q) imp r]; By making a good impression to Mary, Bob gets to know Mary.

Consider Classical Propositional Logic according to which either Bob is sober or Bob is drunk but not both. Thus, we take either p or (non-p) and (A) - (D) for non-logical axioms; take first a theory with p as a non-logical axiom. This theory is satisfiable; indeed, a valuation v such that $v(p) = 1$, $v(q) = v(r) = 0$ satisfies it. By Soundness Theorem, r is not deducible. Similar consequence has a theory in which non-p is a non-logical axiom. Thus, according to Classical Logic, Bob has no chance with Mary at all.

Degree of drunkenness is, however, a typical fuzzy phenomenon. One can be totally tipsy or as sober as judge and everything in between. Consider, therefore, the following fuzzy theory. Assume the set R on fuzzy rules of inference is composed of R_{GMP} and the following schemas

Rule of Bold Disjunction:

$$R_{RBD} \quad : \quad \frac{\alpha, \beta}{(\alpha \text{ or } \beta)}, \frac{a, b}{a \oplus b}$$

Rule of Conjunction:

$$R_{RC} \quad : \quad \frac{\alpha, \beta}{(\alpha \ \overline{\text{and}} \ \beta)}, \frac{a, b}{a \wedge b}$$

It will be an exercise to prove that these schemas are fuzzy rules of inference.

We will need no logical axioms, only the following non-logical axioms:

(A') $T(p \text{ imp non-}q) = 1$,

(B') $T(\text{non-}p \text{ imp non-}q) = 1$,

(C') $T([p\ \overline{\text{and}}\ \text{non-}p] \text{ imp } q) = 1$,

(D') $T([q \text{ or } q] \text{ imp } r) = 1$,

(E) $T(p) = 0.5$,

(F) $T(\text{ non-}p) = 0.5$.

It is easy to see that a valuation v such that $v(p) = v(q) = 0.5$, $v(r) = 1$ satisfies T, hence $\langle$R, $T\rangle$ is consistent. Now we are ready to present the following metaproof for r in $\langle$R, $T\rangle$:

p	,	0.5 ,	*non-logical axiom E*
non-p	,	0.5 ,	*non-logical axiom F*
(p $\overline{\text{and}}$ non-p)	,	0.5 ,	R_{RC}
([p $\overline{\text{and}}$ non-p] imp q)	,	1 ,	*non-logical axiom C'*
q	,	0.5 ,	R_{GMP}
p	,	0.5 ,	*non-logical axiom E*
non-p	,	0.5 ,	*non-logical axiom F*
(p $\overline{\text{and}}$ non-p)	,	0.5 ,	R_{RC}
([p $\overline{\text{and}}$ non-p] imp q)	,	1 ,	*non-logical axiom C'*
q	,	0.5 ,	R_{GMP}
(q or q)	,	1 ,	R_{RBD}
(q or q) imp r	,	1 ,	*non-logical axiom D'*
r	,	1 ,	R_{GMP}

Thus, $\langle \mathtt{R}, T\rangle \vdash_1 r$. Freely speaking, *If Bob is just a bit drunk not to be too shy he will make some kind of positive impression on Mary. Repeating this at least twice (in two different evenings ?) he gets to know Mary.*

3.3 Completeness of Fuzzy Logic

In this section we establish the most important result concerning Fuzzy Propositional Logic; if the set L of values of truth is an injective MV-algebra, in particular, in the unit interval if, and only if $[0,1]$ forms a complete MV-algebra then the semantic consequence operation $\mathcal{C}^{\mathbf{sem}}$ and the syntactic consequence operation $\mathcal{C}^{\mathbf{syn}}$ coincide. In other words, for any fuzzy theory T, for each formula $\alpha \in \mathcal{F}$ and for any value $a \in L$, holds

$$T \vdash_a \alpha \text{ if, and only if } T \models_a \alpha.$$

Since this trivially holds for inconsistent fuzzy theories, we assume the fuzzy theory T under consideration, for now on fixed, is consistent. We start by constructing the *Lindenbaum algebra* of Fuzzy Propositional Logic.

Define on the set $\mathcal{F}$ a binary relation $\preceq$ in the following way:

$$\alpha \preceq \beta \text{ if, and only is } T \vdash_1 (\alpha \texttt{ imp } \beta).$$

By Ax.1, $\preceq$ is reflexive. If $\alpha \preceq \beta$ and $\beta \preceq \gamma$, then $\alpha \preceq \gamma$ (Exercise 1), thus $\preceq$ is transitive, and therefore, a quasi-order. Hence, by defining a binary operation $\sim$ via

$$\alpha \sim \beta \text{ iff } T \vdash_1 (\alpha \texttt{ imp } \beta) \text{ and } T \vdash_1 (\beta \texttt{ imp } \alpha)$$

we obtain an equivalence relation on $\mathcal{F}$. As usual, we denote by $|\alpha|$ the equivalence class defined by α and the set of all equivalence classes by $\mathcal{F}/\sim$. The relation $\sim$ is a congruence with respect to the logical connectives imp and non (Exercise 2). Accordingly, the equation

$$|\alpha| \rightarrow |\beta| = |\alpha \texttt{ imp } \alpha|$$

defines a binary operation $\rightarrow$ on $\mathcal{F}/\sim$, and

$$|\alpha|^* = |\texttt{non-}\alpha|$$

defines a unary operation * on $\mathcal{F}/\sim$. In Exercise 3 one shows that $|\mathbf{0}|$ and $|\mathbf{1}|$ are the least element and the largest element, respectively, in $\mathcal{F}/\sim$ with respect to an order relation given by

$$|\alpha| \leq |\beta| \text{ iff } T \vdash_1 (\alpha \texttt{ imp } \beta).$$

Proposition 97 *For each $\alpha \in \mathcal{F}$, $T \vdash_1 \alpha$ iff $|\alpha| = |\mathbf{1}|$*

Proof. Let $T \vdash_1 \alpha$. By Proposition 96, $T \vdash_1 (\mathbf{1} \texttt{ imp } \alpha)$. Thus, $|\mathbf{1}| \leq |\alpha| \leq |\mathbf{1}|$. Therefore $|\alpha| = |\mathbf{1}|$. Conversely, assume $|\alpha| = |\mathbf{1}|$. Then $T \vdash_1 (\mathbf{1} \texttt{ imp } \alpha)$. Let w be a metaproof for $(\mathbf{1} \texttt{ imp } \alpha)$ with $Val_T(w) = a$. Then we have the following metaproof w' for α:

$$\begin{array}{llll} (\mathbf{1} \texttt{ imp } \alpha) & , \; a & , & \textit{assumption} \\ \mathbf{1} & , \; \mathbf{1} & , & Ax.7 \\ \alpha & , \; a & , & R_{GMP} \end{array}$$

where $Val_T(w') = a$. We conclude $T \vdash_1 \alpha$. □

Proposition 98 *The algebra $\langle \mathcal{F}/\sim, \rightarrow, ^*, |\mathbf{1}| \rangle$ is a Wajsberg algebra.*

Proof. We demonstrate that the Wajsberg axioms (2.1) - (2.4) hold in the set $\mathcal{F}/\sim$. Let $|\alpha|, |\beta|, |\gamma| \in F/\sim$. By Ax.8, we have $|\alpha| \leq |\mathbf{1}| \rightarrow |\alpha|$, and by Ax.9, $|\mathbf{1}| \rightarrow |\alpha| \leq |\alpha|$. Therefore $|\mathbf{1}| \rightarrow |\alpha| = |\alpha|$. Thus, (2.1) holds. (2.2) follows from Ax.2. By Ax.10, we have

$$(|\alpha| \rightarrow |\beta|) \rightarrow |\beta| \leq (|\beta| \rightarrow |\alpha|) \rightarrow |\alpha|$$

and, by changing the roles of $|\alpha|$ and $|\beta|$,

$$(|\beta| \rightarrow |\alpha|) \rightarrow |\alpha| \leq (|\alpha| \rightarrow |\beta|) \rightarrow |\beta|.$$

Therefore $(|\beta| \rightarrow |\alpha|) \rightarrow |\alpha| = (|\alpha| \rightarrow |\beta|) \rightarrow |\beta|$, thus (2.3) holds. Finally, by Ax.11, we have $(|\alpha|^* \rightarrow |\beta|^*) \rightarrow (|\beta| \rightarrow |\alpha|) = |\mathbf{1}|$, whence (2.4) holds. □

We leave as an exercise for the reader to prove that, in the corresponding lattice, the lattice operations $\wedge$ and $\vee$ are defined by

$$|\alpha| \wedge |\beta| = |\alpha \;\overline{\texttt{and}}\; \beta|, \; |\alpha| \vee |\beta| = |\alpha \;\overline{\texttt{or}}\; \beta|,$$

as well as the following

Remark 23 *Assume $h : [\mathcal{F}/\sim] \searrow L$ is an MV-homomorphism and $T \vdash_a \alpha$. Then $h(|\mathbf{a}|) \leq h(|\alpha|)$.*

Proposition 99 *Let T be a consistent fuzzy theory. For each $a, b \in L$, $a \neq b$ iff $|\mathbf{a}| \neq |\mathbf{b}|$.*

Proof. Assume $a \neq b$, in particular, $a \not\leq b$ but $|\mathbf{a}| = |\mathbf{b}|$ (case $b \not\leq a$ is symmetric). Then $T \vdash_1 (\mathbf{a} \texttt{ imp } \mathbf{b})$. If for every metaproof w for $(\mathbf{a} \texttt{ imp } \mathbf{b})$ holds $Val_T(w) \leq a \to b$, then

$$\mathbf{1} = \bigvee\{Val_T(w) \mid w \text{ is a metaproof for } (\mathbf{a} \texttt{ imp } \mathbf{b})\} \leq a \to b,$$

consequently, $a \leq b$, which is not the case. Thus, there exists a metaproof w' for $(\mathbf{a} \texttt{ imp } \mathbf{b})$ such that $Val_T(w') = c$ and $c \not\leq a \to b$, or equivalently, $a \odot c \not\leq b$. Let α be a fixed formula. We have the following metaproof for α:

$\mathbf{a} \texttt{ imp } \mathbf{b}$,	c ,	*assumption*
$\mathbf{a}$,	a ,	$Ax.7$
$\mathbf{b}$,	$c \odot a$,	R_{GMP}
$\mathbf{0}$,	$\mathbf{1}$,	R_{b-CTR}
$\mathbf{0} \texttt{ imp } \alpha$,	$\mathbf{1}$,	$Ax.5$
α ,	$\mathbf{1}$,	R_{GMP}

Thus, $T \vdash_1 \alpha$. This contradict the assumption T is consistent. We conclude $|\mathbf{a}| \neq |\mathbf{b}|$. The converse is trivial. □

Theorem 25 (Completeness Theorem for Fuzzy Propositional Logic) *Any consistent fuzzy theory T is semantically complete.*

Proof. Let $a, b \in L$, α a well-formed formula and $T \vdash_a \alpha$, $T \models_b \alpha$. By Soundness Theorem, $a \leq b$. We construct a truth value function v such that v satisfies T and $v(\alpha) = a$. This shows that $a = b$.

Let $A = \langle A, \to, ^*, |\mathbf{1}|\rangle$ be the subalgebra of the Wajsberg algebra $\mathcal{F}/\sim$ generated by the set $A_\circ = \{|\alpha|\} \cup \{|\mathbf{c}| \mid c \in L\}$. Obviously, A is a Wajsberg algebra such that the operations $\to, ^*$ coincide with those of $\mathcal{F}/\sim$. Define a mapping $h_\circ : A_\circ \searrow L$ by $h_\circ(|\alpha|) = a$ and $h_\circ(|\mathbf{c}|) = c$ for each $c \in L$. The mapping $h_\circ$ is well-defined and can be extended in natural way on A and, since L is injective, on the whole Wajsberg algebra $\mathcal{F}/\sim$. Denote the extended map by h. Let k be the *natural homomorphism*, i.e. $k(\beta) = |\beta|$ for any $\beta \in \mathcal{F}$. Then the mapping $h \circ k$: $\mathcal{F} \searrow L$ is a truth value function. Indeed, $h \circ k(\beta)$ is defined for all $\beta \in \mathcal{F}$, $h \circ k(\mathbf{c}) = h(|\mathbf{c}|) = h_\circ(|\mathbf{c}|) = c$ for each $c \in L$ and, for each $\beta, \gamma \in \mathcal{F}$

$$\begin{aligned} h \circ k(\beta \texttt{ imp } \gamma) &= h(|\beta \texttt{ imp } \gamma|) \\ &= h(|\beta|) \to h(|\gamma|) \\ &= h \circ k(\beta) \to h \circ k(\gamma). \end{aligned}$$

Assume $\beta \in \mathcal{F}$ is a non-logical axiom of T and $T(\beta) = c$, $T \vdash_d \beta$. Then $c \leq d = h(|\mathbf{d}|) \leq h(|\beta|) = h \circ k(\beta)$, hence $T(\beta) \leq h \circ k(\beta)$, accordingly, $h \circ k$

satisfies T. Finally, $h \circ k(\alpha) = h(|\alpha|) = h_\circ(|\alpha|) = a$. Consequently, $h \circ k$ is the truth value function we are looking for. □

Corollary Any satisfiable fuzzy theory T is complete.

In Classical Propositional Logic holds the following *Deduction Theorem*

$$\text{If } \alpha \vdash \beta \text{ then } \vdash (\alpha \texttt{ imp } \beta).$$

In Fuzzy Propositional Logic the situation is different as we shall now see. Consider Łukasiewicz valued fuzzy theory T with one non-logical axiom $T(\mathbf{a} \texttt{ imp } [\mathbf{a} \texttt{ and } p]) = 1$, where p is a propositional variable and $a = \frac{1}{4}$. For any valuation v which satisfies T holds $v(\mathbf{a} \texttt{ imp } [\mathbf{a} \texttt{ and } p]) = 1$, or equivalently, $v(\mathbf{a}) \leq v(\mathbf{a}) \odot v(p)$ implying $v(p) = 1$. Thus, $T \models_1 p$ and by Completeness Theorem, $T \vdash_1 p$. However, for a valuation v such that $v(p) = 0.45$ it holds that $v(\{\mathbf{a} \texttt{ imp } [\mathbf{a} \texttt{ and } p]\} \texttt{ imp } p) = 0.70$. Therefore

$$\not\models_1 (\{\mathbf{a} \texttt{ imp } [\mathbf{a} \texttt{ and } p]\} \texttt{ imp } p)$$

whence

$$\not\vdash_1 (\{\mathbf{a} \texttt{ imp } [\mathbf{a} \texttt{ and } p]\} \texttt{ imp } p).$$

Thus, by taking $\alpha = (\mathbf{a} \texttt{ imp } [\mathbf{a} \texttt{ and } p])$, $\beta = p$ we have a case such that

$$\alpha \vdash_1 \beta \text{ holds, while } \not\vdash_1 (\alpha \texttt{ imp } \beta).$$

Fuzzy rules of inference R_{GMP}, R_{a-CTR}, R_{a-LR} and R_{RBC} are sufficient to establish the Completeness Theorem of Fuzzy Predicate Logic. Since fuzzy rules of inference are sound, introducing new fuzzy rules of inference into a fuzzy theory does not have any effect on the degree of deduction of any formula α. Metaproofs, however, may turn to be easier.

Proposition 100 *In complete MV-algebra valued Fuzzy Propositional Logic, the following schemas are fuzzy rules of inference.*

Generalized Modus Tollendo Tollens:

$$R_{GMTT} \quad : \quad \frac{\texttt{non-}\beta, (\alpha \texttt{ imp } \beta)}{\texttt{non-}\alpha}, \quad \frac{a, b}{a \odot b}$$

Generalized Hypothetical Syllogism:

$$R_{GHS} \quad : \quad \frac{(\alpha \texttt{ imp } \beta), (\beta \texttt{ imp } \gamma)}{(\alpha \texttt{ imp } \gamma)}, \quad \frac{a, b}{a \odot b}$$

Generalized Commutative Law 1:

$$R_{GCL1} \quad : \quad \frac{(\alpha \texttt{ and } \beta)}{(\beta \texttt{ and } \alpha)}, \frac{a}{a}$$

Generalized Commutative Law 2:

$$R_{GCL2} \quad : \quad \frac{(\alpha \texttt{ or } \beta)}{(\beta \texttt{ or } \alpha)}, \frac{a}{a}$$

Generalized Equivalence Law 1:

$$R_{GEL1} \quad : \quad \frac{(\alpha \texttt{ equiv } \beta)}{(\alpha \texttt{ imp } \beta)}, \frac{a}{a}$$

Generalized Equivalence Law 2:

$$R_{GEL2} \quad : \quad \frac{(\alpha \texttt{ equiv } \beta)}{(\beta \texttt{ imp } \alpha)}, \frac{a}{a}$$

Generalized Equivalence Law 3:

$$R_{GEL3} \quad : \quad \frac{(\alpha \texttt{ imp } \beta), (\beta \texttt{ imp } \alpha)}{(\alpha \texttt{ equiv } \beta)}, \frac{a, b}{a \wedge b}$$

Generalized Simplification Law 1:

$$R_{GS1} \quad : \quad \frac{(\alpha \texttt{ and } \beta)}{\alpha}, \frac{a}{a}$$

Generalized Simplification Law 2:

$$R_{GS2} \quad : \quad \frac{(\alpha \texttt{ and } \beta)}{\beta}, \frac{a}{a}$$

Generalized Rule of Introduction of Double Negation:

$$R_{GIDN} \quad : \quad \frac{\alpha}{\texttt{non-(non-}\alpha\texttt{)}}, \frac{a}{a}$$

Generalized Rule of Elimination of Double Negation:

$$R_{GEDN} \quad : \quad \frac{\texttt{non-(non-}\alpha\texttt{)}}{\alpha}, \frac{a}{a}$$

Generalized De Morgan Law 1:

$$R_{GDeM1} \quad : \quad \frac{(\texttt{non-}\alpha) \texttt{ and } (\texttt{non-}\beta)}{\texttt{non-}(\alpha \texttt{ or } \beta)}, \frac{a}{a}$$

Generalized De Morgan Law 2:

$$R_{GDeM2} \quad : \quad \frac{\texttt{non-}(\alpha \texttt{ or } \beta)}{(\texttt{non-}\alpha) \texttt{ and } (\texttt{non-}\beta)}, \frac{a}{a}$$

Generalized De Morgan Law 3:

$$R_{GDeM3} \quad : \quad \frac{(\texttt{non-}\alpha) \texttt{ or } (\texttt{non-}\beta)}{\texttt{non-}(\alpha \texttt{ and } \beta)}, \frac{a}{a}$$

Generalized De Morgan Law 4:

$$R_{GDeM4} \quad : \quad \frac{\texttt{non-}(\alpha \texttt{ and } \beta)}{(\texttt{non-}\alpha) \texttt{ or } (\texttt{non-}\beta)}, \frac{a}{a}$$

Generalized Addition Law:

$$R_{GAL} \quad : \quad \frac{\alpha}{(\alpha \texttt{ or } \beta)}, \frac{a}{a}$$

Generalized Modus Tollendo Ponens:

$$R_{GMTP} \quad : \quad \frac{\texttt{non-}\beta, (\alpha \texttt{ or } \beta)}{\alpha}, \frac{a, b}{a \odot b}$$

Generalized Disjunctive Syllogism:

$$R_{GDS} \quad : \quad \frac{(\alpha \texttt{ or } \beta), (\alpha \texttt{ imp } \gamma), (\beta \texttt{ imp } \delta)}{(\gamma \texttt{ or } \delta)}, \frac{a, b, c}{a \odot b \odot c}$$

Generalized Rule of Introduction of Implication:

$$R_{II} \quad : \quad \frac{(\texttt{non-}\alpha \texttt{ or } \beta)}{(\alpha \texttt{ imp } \beta)}, \frac{a}{a}$$

Generalized Rule of Elimination of Implication:

$$R_{EI} \quad : \quad \frac{(\alpha \ \texttt{imp} \ \beta)}{(\texttt{non-}\alpha \ \texttt{or} \ \beta)}, \ \frac{a}{a}$$

Proof. Semi-continuity of the schemas R_{GMTT}, R_{GHS}, R_{GMTP}, as well as of the schema R_{GDS} follows essentially from the fact that

$$r^{\texttt{sem}}(a, \bigvee_{i\in\Gamma} b_i) = a \odot (\bigvee_{i\in\Gamma} b_i) = \bigvee_{i\in\Gamma}(a \odot b_i) = \bigvee_{i\in\Gamma} r^{\texttt{sem}}(a, b_i),$$

and, for the schema R_{GEL3}, by

$$r^{\texttt{sem}}(a, \bigvee_{i\in\Gamma} b_i) = a \wedge (\bigvee_{i\in\Gamma} b_i) = \bigvee_{i\in\Gamma}(a \wedge b_i) = \bigvee_{i\in\Gamma} r^{\texttt{sem}}(a, b_i).$$

The other schemas are trivially semi-continuous. Soundness of most of these schemas is a direct consequence from properties on valuation. We establish here soundness of R_{GMTT} and R_{GMTP}. R_{GDS} is left as an exercise for the reader. Assume v is a valuation. Then

$$\begin{aligned} r^{\texttt{sem}}(v(\texttt{non-}\beta), v(\alpha \ \texttt{imp} \ \beta)) &= v(\texttt{non-}\beta) \odot v(\alpha \ \texttt{imp} \ \beta) \\ &= v(\beta)^* \odot [v(\alpha) \to v(\beta)] \\ &= [v(\alpha) \to v(\beta)] \odot [v(\beta) \to \mathbf{0}] \\ &\leq v(\alpha) \to \mathbf{0} \\ &= v(\texttt{non-}\alpha) \\ &= v(r^{\texttt{syn}}(\texttt{non-}\beta, [\alpha \ \texttt{imp} \ \beta])). \end{aligned}$$

Thus, R_{GMTT} is sound. For R_{GMTP} we have

$$\begin{aligned} r^{\texttt{sem}}(v(\texttt{non-}\beta), v(\alpha \ \texttt{or} \ \beta)) &= v(\texttt{non-}\beta) \odot v(\alpha \ \texttt{or} \ \beta) \\ &= v(\beta)^* \odot [v(\alpha) \oplus v(\beta)] \\ &= v(\beta)^* \odot [v(\beta)^* \to v(\alpha)] \\ &\leq v(\alpha) \\ &= v(r^{\texttt{syn}}(\texttt{non-}\beta, [\alpha \ \texttt{or} \ \beta])). \end{aligned}$$

□

Example 1 Recall Example 3.1.1 where we studied the following

Assuming winter is long implies either winter is not cold or damages caused by floods are not serious, then insurance business is profitable.

This was expressed formally by $\{[p \ \texttt{imp} \ (\texttt{non-}q \ \overline{\texttt{or}} \ \texttt{non-}r)] \ \texttt{imp} \ s\}$. In particular, we set $T(p) = 0.2$, $T(q) = 0.8$, $T(r) = 0.1$, $T(s) = 0.8$, $T(\beta) = 0$ elsewhere. Then we reasoned

$$T \models_{0.8} \{[p \ \texttt{imp} \ (\texttt{non} - q \ \overline{\texttt{or}} \ \texttt{non-}r)] \ \texttt{imp} \ s\}.$$

Now we are looking for metaproofs for this same formula. For example, we have w_1:

s	, 0.8 ,	*non-logical axiom*
(non-β or s)	, 0.8 ,	R_{GAL}
(β imp s)	, 0.8 ,	R_{II}

where β stands for [p imp (non $-$ q $\overline{\text{or}}$ non-r)]. Thus, $Val_T(w_1) = 0.8$. Moreover, we have w_2:

[s imp (β imp s)]	, 1 ,	$Ax.8$
s	, 0.8 ,	*non-logical axiom*
(β imp s)	, 0.8 ,	R_{GMP}

Again, $Val_T(w_2) = 0.8$. Since the fuzzy theory T is satisfiable it is complete. Thus, $T \vdash_{0.8}$ {[p imp (non $-$ q $\overline{\text{or}}$ non-r)] imp s}. Freely expressed, *The statement is, under given conditions, mostly true.*

Example 2 In Exercises 3.1.2 we introduced a Łukasiewicz valued fuzzy theory T such that
T([(p or q) imp r]) = 1, T(r imp (s or t)) = 0.9, T(t imp w) = 1, T(non-s and non-w) = 0.8, T(non-p) = 1, where p stands for *Wages rise,* q stands for *Prices rise,* r stand for *There is inflation,* s stands for *The Government stops inflation,* t stands for *People suffer,* w stands for *The Government loses its popularity.* In Exercise 3.1.4 we showed that $\mathcal{C}^{\text{sem}}(T)$(non-$p$) = 0.7. Now we are looking for a metaproof for the formula non-p. We find the following:

(1)	[(p or q) imp r]	, 1 ,	*non-logical axiom*
(2)	[r imp (s or t)]	, 0.9 ,	*non-logical axiom*
(3)	(t imp w)	, 0.8 ,	*non-logical axiom*
(4)	(non-s and non-w)	, 1 ,	*non-logical axiom*
(5)	non-w	, 1 ,	(4) , R_{GS2}
(6)	non-s	, 1 ,	(4) , R_{GS1}
(7)	non-t	, 0.8 ,	(5) , (3) , R_{GMTT}
(8)	(non-s and non-t)	, 0.8 ,	(6) , (7) , R_{RBC}
(9)	non-(s or t)	, 0.8 ,	(8) , R_{GDeM1}
(10)	non-r	, 0.7 ,	(9) , (2) , R_{GMTT}
(11)	non-(p or q)	, 0.7 ,	(10) , (1) , R_{GMTT}
(12)	(non-p and non-q)	, 0.7 ,	(11) , R_{GDeM2}
(13)	non-p	, 0.7 ,	(12) , R_{GS1}

We conclude that $\mathcal{C}^{\text{syn}}(T)$(non-$p$) = 0.7, a fact which, of course, follows also by Completeness Theorem.

Evidently, the longer a deduction from partial true premises is, the more unsure is the truth of the conclusion. Due to non-idempotency of the operations $\odot$ and $\oplus$ in MV-algebras, Fuzzy Propositional Logic obeys this property as we already have see. Consider also the following

Example 3 *It is* mostly true *that if a Finn understands the Swedish spoken in Finland then he understands the Swedish spoken in Stockholm, and it is* mostly true *that if one understands the Swedish spoken in Stockholm, then one understands the dialect of Southern Sweden, as well as it is* mostly true *that the one who understands the dialect of Southern Sweden will understand Danish. Assume Paavo understands* more or less *the Swedish spoken in Finland. Under these premises, at what degree Paavo understands Danish?*

We define a Łukasiewicz valued fuzzy theory T such that $T(p) = 0.7$, $T(p \texttt{ imp } q) = T(q \texttt{ imp } r) = T(r \texttt{ imp } s) = 0.9$, where p stands for *Paavo understands the Swedish spoken in Finland,* q stands for *Paavo understands the Swedish spoken in Stockholm,* r stands for *Paavo understands the dialect of Southern Sweden,* and s stands for *Paavo understands Danish.*

This fuzzy theory is satisfiable, indeed, a valuation v such that $v(p) = 1$, $v(q) = 0.9$, $v(r) = 0.8$, $v(s) = 0.7$ satisfies T. We have the following metaproof for the formula s:

$$\begin{array}{llll}
p & , & 0.7 & , \quad \textit{non-logical axiom} \\
(p \texttt{ or } q) & , & 0.9 & , \quad \textit{non-logical axiom} \\
q & , & 0.6 & , \quad R_{GMP} \\
(q \texttt{ or } r) & , & 0.9 & , \quad \textit{non-logical axiom} \\
r & , & 0.5 & , \quad R_{GMP} \\
(r \texttt{ or } s) & , & 0.9 & , \quad \textit{non-logical axiom} \\
s & , & 0.4 & , \quad R_{GMP}
\end{array}$$

Thus, *Paavo understands Danish* is at least 0.4-true. It is also at most 0.4-true as a valuation v such that $v(p) = 0.7$, $v(q) = 0.6$, $v(r) = 0.5$, $v(s) = 0.4$ also satisfies T. Since $C^{syn}(T)(s) = 0.4$, the right conclusion would be *Paavo understands some basic Danish.*

A conclusion of partially true statements may sometimes have a greater value of truth than any of the premises. Consider

Example 4 *A technical system needs control if any of its four subsystems is performing badly. Assume all the four subsystems are performing* more or less *well. Does the system need technical control?*

Define a Łukasiewicz valued fuzzy theory T by setting $T([p_1 \texttt{ or } p_2 \texttt{ or } p_3 \texttt{ or } p_4] \texttt{ imp } q) = 1$, $T(p_i) = 0.3$ for $i = 1, \cdots, 4$ where p_i stands for *The subsystem i is performing badly,* and q stands for *The system needs technical control.* This fuzzy theory is satisfiable, indeed, a valuation v such that $v(p_i) = v(q) = 1$ for $i = 1, \cdots, 4$ satisfies T. We have the following metaproof for the formula q:

p_1	, 0.3 ,	*non-logical axiom*
p_2	, 0.3 ,	*non-logical axiom*
p_1 or p_2	, 0.6 ,	R_{RBD}
p_3	, 0.3 ,	*non-logical axiom*
p_1 or p_2 or p_3	, 0.9 ,	R_{RBD}
p_4	, 0.3 ,	*non-logical axiom*
p_1 or p_2 or p_3 or p_4	, 1 ,	R_{RBD}
$[p_1$ or p_2 or p_3 or $p_4]$ imp q	, 1 ,	*non-logical axiom*
q	, 1 ,	R_{GMP}

Our conclusion is *The system needs a technical control.*

3.4 Exercises

3.1. Semantics of Fuzzy Propositional Logic

Exercise 1 Formalize the following statements of natural language.
(a) Mary is blond.
(b) If Mary is right, then Bob is wrong.
(c) Bob made a mistake or there is no ball tonight.
(d) If it is raining Bob cannot go to ball by walk.
(e) If I'll work through the whole night I'll be tired.
(f) Either it is raining and the Sun is shining or we cannot see the rainbow.
(g) It is raining or sunny but it is not windy.

Exercise 2 Formalize the following statements of natural language.
(a) If wages rise or prices rise there will be inflation.
(b) If there will be inflation, then the Government has to stop it or people will suffer.
(c) If people will suffer the Government will lose its popularity.
(d) The Government will not stop the inflation and the Government will not lose its popularity.
(e) Wages will not rise.

Exercise 3 Prove equations (3.4) and (3.5).

Exercise 4 Let α, β, γ, δ, ϵ be formulas (a), (b), (c), (d), (e) of Exercise 2, respectively. Define a fuzzy set of formulas T by $T(\alpha) = 1$, $T(\beta) = 0.9$, $T(\gamma) = 0.8$ and $T(\delta) = 1$. Calculate the degree of validity of the formula ϵ in Łukasiewicz valued Fuzzy Propositional Logic. Use
(a) **and** and **or** connectives,
(b) $\overline{\text{and}}$ and $\overline{\text{or}}$ connectives.

Exercise 5 Let v be a valuation such that, for each formula $\alpha \in \mathcal{F}$, $v(\alpha) \in \{0, 1\}$. Show that

(a) v satisfies the truth tables of Classical Propositional Logic,

(b) $v(\alpha \texttt{ and } \beta) = v(\alpha \overline{\texttt{and}} \beta)$, $v(\alpha \texttt{ or } \beta) = v(\alpha \overline{\texttt{or}} \beta)$.

Exercise 6 Fill the following instances of truth tables of Łukasiewicz valued Fuzzy Propositional Logic

$v(\beta)$

	$v(\alpha$ **and** $\beta)$	0.0	0.2	0.4	0.6	0.8	1.0
	0.0						
	0.2						
$v(\alpha)$	0.4						
	0.6						
	0.8						
	1.0						

$v(\beta)$

	$v(\alpha$ `or` $\beta)$	0.0	0.2	0.4	0.6	0.8	1.0
	0.0						
	0.2						
$v(\alpha)$	0.4						
	0.6						
	0.8						
	1.0						

$v(\beta)$

	$v(\alpha$ `imp` $\beta)$	0.0	0.2	0.4	0.6	0.8	1.0
	0.0						
	0.2						
$v(\alpha)$	0.4						
	0.6						
	0.8						
	1.0						

Exercise 7 Show that definition (3.7) generalizes tautology of Classical Propositional Logic.

Exercise 8 Prove equations (3.9) - (3.19).

Exercise 9 Let α stand for $\{p \texttt{ imp } [q \texttt{ and non-}(r \texttt{ or non-}s)]\}$. Construct a formula β such that β contains only logical connective `imp` and, for any valuation v, $v(\alpha) = v(\beta)$. What is the main `imp`-connective in β?

Exercise 10 Let xor be a logical connective called *exclusive or* and defined by an abbreviation

$(\alpha \texttt{ xor } \beta) = (\alpha \texttt{ or } \beta)\ \overline{\texttt{and}}\ (\alpha \texttt{ imp non-}\beta)\ \overline{\texttt{and}}\ (\beta \texttt{ imp non-}\alpha)$.

(a) Express $v(\alpha \texttt{ xor } \beta)$ by means of $v(\alpha)$, $v(\beta)$ and the MV-algebra operations $\oplus$, $\wedge$ and *.

(b) Fill the following instance of truth table of Łukasiewicz valued Fuzzy Propositional Logic for the logical connective xor

		$v(\beta)$					
	$v(\alpha \texttt{ xor } \beta)$	0.0	0.2	0.4	0.6	0.8	1.0
	0.0						
	0.2						
$v(\alpha)$	0.4						
	0.6						
	0.8						
	1.0						

(c) Show that, for any valuation v, any formulas α, β,
$v(\alpha \texttt{ xor } \beta) = v(\beta \texttt{ xor } \alpha)$, $v(\alpha \texttt{ xor } \beta) = v([\alpha \texttt{ or } \beta]\ \overline{\texttt{and}}\ \texttt{non-}[\alpha \texttt{ and } \beta])$.

Exercise 11 Formalize the following statements of natural language.
(a) Either Bob reads for the examination or goes for beer but not both.
(b) If Bob reads for the examination then Paavo will pass the examination.
(c) If Bob wants to meet girls he cannot read for the examination.
(d) Bob wants to meet girls and have beer.
(e) Bob passes the examination.

Exercise 12 Let α, β, γ, δ, s be formulas (a), (b), (c), (d), (e) of Exercise 11, respectively. Define a fuzzy set of formulas T by $T(\alpha) = 0.8$, $T(\beta) = 0.8$, $T(\gamma) = 0.7$ and $T(\delta) = 1$. Calculate the degree of validity of the formula s in Łukasiewicz valued Fuzzy Propositional Logic.

Exercise 13 Assumptions as in Exercise 12, calculate $C^{\texttt{sem}}(T)(\texttt{non}-\alpha)$. What do you realize?

3.2. Axiom System for Fuzzy Logic

Exercise 1 Construct a unary operation $R : [0,1] \searrow [0,1]$ which is isotone but not semi-continuous.

Exercise 2 Consider a finite chain L as a lattice. Does there exist a unary operation $R : L \searrow L$ which is isotone but not semi-continuous?

Exercise 3 Consider complete MV-algebra valued fuzzy theory $\langle \mathtt{A}, \mathtt{R}, T, \rangle$ containing
(a) the fuzzy rule of inference R_{GMP}. Prove that

$$C^{\mathtt{syn}\langle \mathtt{A},\mathtt{R}\rangle}(T)(\alpha) \odot C^{\mathtt{syn}\langle \mathtt{A},\mathtt{R}\rangle}(T)(\alpha \ \mathtt{imp}\ \beta) \leq C^{\mathtt{syn}\langle \mathtt{A},\mathtt{R}\rangle}(T)(\beta).$$

(b) the fuzzy rules of inference $R_{\mathbf{a}-LR}$ and Ax.7. Prove that

$$a \to C^{\mathtt{syn}\langle \mathtt{A},\mathtt{R}\rangle}(T)(\alpha) \leq C^{\mathtt{syn}\langle \mathtt{A},\mathtt{R}\rangle}(T)(\mathbf{a}\ \mathtt{imp}\ \alpha).$$

(c) the fuzzy rule of inference R_{RBC}. Prove that

$$C^{\mathtt{syn}\langle \mathtt{A},\mathtt{R}\rangle}(T)(\alpha) \odot C^{\mathtt{syn}\langle \mathtt{A},\mathtt{R}\rangle}(T)(\beta) \leq C^{\mathtt{syn}\langle \mathtt{A},\mathtt{R}\rangle}(T)(\alpha \ \mathbf{and}\ \beta).$$

Exercise 4 A function $m : [0,1]^2 \searrow [0,1]$ is called *Modus Ponens generating function* [1] if it satisfies, for all valuations v, all formulas α, β,
(MP1) $m(v(\alpha), v(\alpha \ \mathtt{imp}\ \beta)) \leq v(\beta)$,
(MP2) $m(1,1) = 0$,
(MP3) $m(0,a) = 1$,
(MP4) if $a \leq b$ then $m(a,c) \leq m(b,c)$.
Show that the product operation $\odot$ of any residuated lattice defined on the unit interval is a Modus Ponens generating function.

Exercise 5 A continuous function $n : [0,1] \searrow [0,1]$ is called *negation generating function* if it is strictly decreasing and satisfies, for all $a \in [0,1]$,
(N1) $n(n(a)) = a$,
(N2) $n(1) = 0$,
(N3) $n(0) = 1$.
Show that the complement operation * of any MV-algebra defined on the unit interval is a negation generating function.

Exercise 6 A function $i : [0,1]^2 \searrow [0,1]$ is called *implication generating function* if it satisfies, for each $a, b, c \in [0,1]$,
(I1) if $a \leq b$ then $i(b,c) \leq i(a,c)$,
(I2) if $a \leq b$ then $i(c,a) \leq i(c,b)$,
(I3) $i(0,a) = 1$,
(I4) $i(1,a) = a$,
(I5) $i(a, i(b,c) = i(b, i(a,c)$.
Show that the residual operation $\to$ of any residuated lattice defined on the unit interval is an implication generating function.

Exercise 7 A function $t : [0,1]^2 \searrow [0,1]$ is called *Modus Tollens generating function* if it satisfies, for each valuations v, all formulas α, β,

[1] Definitions in Exercises 4 - 9 are due to Trillas and Valverde.

(MT1) $t(v(\texttt{non-}\beta), v(\alpha \texttt{ imp } \beta)) \leq v(\texttt{non-}\alpha)$,
(MT2) $t(1,1) = 1$,
(MT3) $t(0,a) = 0$, (MP4) if $a \leq b$ then $t(a,c) \leq t(b,c)$.
Show that the product operation $\odot$ of any residuated lattice defined on the unit interval is a Modus Tollens generating function.

Exercise 8 A negation function n and an implication function i are called *compatible* if, for each $a, b, c \in [0,1]$,

$$i(a,b) = i(n(b), n(a)).$$

Moreover, an implication function i is called an *S-implication* if there exists a negation function n such that n and i are compatible. Show that the residual operation $\rightarrow$ of any MV-algebra defined on the unit interval is an S-implication.

Exercise 9 A function $R : [0,1]^2 \searrow [0,1]$ is called *R-implication* if there exists a T-norm t (see Exercise 1.3.16), where t is a continuous function, and such that for each $a, b, c \in [0,1]$,

$$R(a,b) = \bigvee \{c \mid t(a,c) \leq b\}.$$

Show that the residual operation $\rightarrow$ of any complete MV-algebra defined on the unit interval is an R-implication.

Exercise 10 Prove that
(a) Rule of Bold Disjunction is a fuzzy rule of inference in any complete MV-algebra valued Fuzzy Propositional Logic,
(b) Rule of Conjunction is a fuzzy rule of inference in any residuated lattice valued Fuzzy Propositional Logic.

Exercise 11 Show that *Rule of Exclusive or*, defined by

$$R_{XOR} \quad : \quad \frac{(\alpha \texttt{ xor } \beta),\ \texttt{non-}\alpha}{\beta} \quad , \quad \frac{a, b}{a \odot b}$$

is a fuzzy rule of inference in any complete MV-algebra valued Fuzzy Propositional Logic.

Exercise 12 Recall poor Bob (Exercise 3.1.11) who decided to have fun with girls and to drink beer, thus, did not pass his examination. Assume he has now improved his habits. He goes for just one beer and comes then quickly home to read for examination. Formally we express this by writing $T(p \texttt{ xor } q) = 1$, $T(p \texttt{ imp } s) = 1$, $T(\texttt{non} - q) = 0.9$, where
p stands for *Bob reads for the examination*, q stand for *Bob goes for beer*, s stand for *Bob passes the examination.*

Construct a Łukasiewicz valued fuzzy theory and study the degree of deduction of the formula *Bob passes the examination.* How do You interpret the result?

3.3. Completeness of Fuzzy Logic

Exercise 1 Define on the set $\mathcal{F}$ of well-formed formulas a relation $\preceq$ via $\alpha \preceq \beta$ if, and only if $T \vdash_1 (\alpha \texttt{ imp } \beta)$.

Show that $\preceq$ is transitive.

Exercise 2 Prove that an equivalence relation $\sim$, defined by

$$\alpha \sim \beta \text{ iff } T \vdash_1 (\alpha \texttt{ imp } \beta) \text{ and } T \vdash_1 (\beta \texttt{ imp } \alpha),$$

is a congruence with respect to the logical connective `imp` and `non`.

Exercise 3 Prove that $|\mathbf{0}|$ and $|\mathbf{1}|$ are the least element and the greatest element in $\mathcal{F}/\sim$ with respect to an order relation given by $|\alpha| \leq |\beta|$ iff $T \vdash_1 (\alpha \texttt{ imp } \beta)$.

Exercise 4 Assume $h : [\mathcal{F}/\sim] \searrow L$ is an MV-homomorphism and $T \vdash_a \alpha$. Verify $h(|\mathbf{a}|) \leq h(|\alpha|)$.

Exercise 5 Prove, in the MV-algebra $[\mathcal{F}/\sim]$, that $|\alpha| \wedge |\beta| = |\alpha \ \overline{\textbf{and}} \ \beta|$, $|\alpha| \vee |\beta| = |\alpha \ \overline{\texttt{or}} \ \beta|$

Exercise 6 Prove that Generalized Disjunctive Syllogism is sound.

Exercise 7 Nervousness usually ensues from precipitation, and bad eating habits usually cause peritonitis. Gastric ulcer, in turn, is always due to nervousness or peritonitis. For an old person gastric ulcer is a fatal disease. Assume a middle-aged person is regularly in a hurry and has unwholesome diet. Will he contract gastric ulcer? If so, is the illness fatal? Formulate a Łukasiewicz valued fuzzy theory and answer to the questions by means of truth value functions as well as by means of metaproofs. Pay a special tension on how to express *usually* with respect to *always* and *middle-aged* with respect to *old.*

Exercise 8 If Mary, a single white female, is wearing basic black and high-heeled shoes, she looks tempting. If Mary is wearing jacket and low-heeled shoes, she looks academic. If Mary wears pealed cap or jeans or tennis shoes, she is relaxed. Mary would like to make an impression on Bob. If Mary manages to make an impression on Bob the story has a happy end.

When Bob is sober he likes academic women but when Bob is drunk he cannot stand them and is crazy with women who look tempting. Relaxed women Bob likes always. This evening there is ball and Mary wears jeans, jacket and high-heeled shoes. Bob, in turn, is just a bit tipsy. Does the story have a happy end? Study the case in Łukasiewicz valued Fuzzy Propositional Logic.

Chapter 4

Fuzzy Relations

Fuzzy relations, that is [0,1]-valued binary relations were introduced by Zadeh in 1965. He also established their sum-min composition. Later other kind of compositions between fuzzy relations has been introduced. In this last chapter we study the theory of fuzzy relations from a general algebraic point of view. We show their connection to so-called *approximate reasoning* familiar in fuzzy systems context. We conclude our study on fuzzy similarity relation, which is fundamental in many-valued reasoning. We let the set of values of degree of relationship be a complete generalized residuated lattice $\langle L, \leq, \wedge, \vee, \mu, f, g\rangle$ (see Section 1.3). This allows any kind of sup-μ composition between fuzzy relations; the binary operation μ may be any kind of complete MV-algebra product, the Gaines product, the min-operation, a pseudo Boolean algebraic $\wedge$-operation, any t-norm, etc.

4.1 Solvability of Fuzzy Relation Equations

In this section we focus on the solvability of fuzzy relation equation.

Let U, V be sets. A binary[1] *fuzzy relation* R on $U \times V$ is a fuzzy set on $U \times V$, i.e. R is a function $R : U \times V \searrow L$, where L is a complete generalized residuated lattice. In such a case we will write $R \underset{\sim}{\subset} U \times V$. The *inverse* of R, denoted by R^{-1}, is the fuzzy relation on $V \times U$ defined by $R^{-1}\langle v, u\rangle = R\langle u, v\rangle$. The *identical* fuzzy relation I is defined by

$$I\langle u, v\rangle = \begin{cases} 1 & \text{for } u = v \\ 0 & \text{otherwise.} \end{cases}$$

[1]We consider here only binary fuzzy relations; many of the concepts we introduce can be easily generalized to n-ary fuzzy relations, most of the applications, however, are based on binary ones.

Definition 31 *Let L be a complete generalized residuated lattice and let $R \lessapprox U \times V$, $S \lessapprox V \times W$ and $T \lessapprox U \times W$ be binary fuzzy relations with values in L. The μ-*composition *of R and S is a fuzzy relation $R\mu S \lessapprox U \times W$ defined by*

$$R\mu S\langle u, w\rangle = \bigvee\nolimits_{v\in V} \mu(R\langle u, v\rangle, S\langle v, w\rangle) \text{ for each } u \in U,\ w \in W.$$

Fuzzy relations $H_1(S,T) \lessapprox U \times V$ and $H_2(R,T) \lessapprox V \times W$ are defined by

$$H_1(S,T)\langle u, v\rangle = \bigwedge\nolimits_{w\in W} f(S\langle v, w\rangle, T\langle u, w\rangle) \text{ for each } u \in U,\ v \in V,$$

and

$$H_2(R,T)\langle v, w\rangle = \bigwedge\nolimits_{u\in U} g(R\langle u, v\rangle, T\langle u, w\rangle) \text{ for each } v \in V,\ w \in W.$$

Remark 24 *The operation μ on fuzzy relations is associative and isotone with respect to the natural order relation $R \leq Q$ iff $R\langle u, v\rangle \leq Q\langle u, v\rangle$ for each $u \in U$, $v \in V$.*

Proof. Exercise.

Remark 25 *In the case of μ commutative we have*

$$(R\mu S)^{-1} = S^{-1}\mu R^{-1} \text{ and } H_2(R,T) = H_1(R^{-1}, T^{-1})$$

Proof. Exercise.

Theorem 26 *$R\mu S \leq T$ iff $R \leq H_1(S,T)$ iff $S \leq H_2(R,T)$.*

Proof. $R\mu S \leq T$ iff

$$\bigvee_{v\in V} \mu(R\langle u, v\rangle, S\langle v, w\rangle) \leq T\langle u, w\rangle \text{ for each } u \in U, w \in W \tag{4.1}$$

$$\text{iff } \mu(R\langle u, v\rangle, S\langle v, w\rangle) \leq T\langle u, w\rangle \text{ for each } u \in U, v \in V w \in W$$

$$\text{iff } R\langle u, v\rangle \leq f(S\langle v, w\rangle, T\langle u, w\rangle) \text{ for each } u \in U, v \in V, w \in W$$

$$\text{iff } R\langle u, v\rangle \leq \bigwedge_{w\in W} f(S\langle v, w\rangle, T\langle u, w\rangle) \text{ for each } u \in U, v \in V$$

iff $R \leq H_1(S,T)$. The (in-)equality (4.1) is also equivalent to

$$S\langle v, w\rangle \leq g(R\langle u, v\rangle, T\langle u, w\rangle) \text{ for each } u \in U, v \in V, w \in W$$

$$\text{iff } S\langle v, w\rangle \leq \bigwedge_{u\in U} g(R\langle u, v\rangle, T\langle u, w\rangle) \text{ for each } v \in V, w \in W$$

iff $S \leq H_2(R,T)$. The proof is complete. □

Since $H_1(S,T) \leq H_1(S,T)$ we obtain

Corollary 1 *$H_1(S,T)\mu S \leq T$.*

and similarly, since $H_2(R,T) \leq H_2(R,T)$

Corollary 2 *$R\mu H_2(R,T)\mu S \leq T$.*

Theorem 27 *The fuzzy relation equation*

$$X\mu S = T \tag{4.2}$$

has a solution X iff $H_1(S,T)$ is a solution. If a solution exists, then $H_1(S,T)$ is the largest one.

Proof. If $H_1(S,T)$ is a solution, then (4.2) has a solution. Let R be a solution of (4.2). Then

$$R \leq H_1(S,T) \tag{4.3}$$

and, by the isotonicity of μ, we have

$$T = R\mu S \leq H_1(S,T)\mu S \leq T \text{ (by Corollary 1),}$$

i.e. $H_1(S,T)$ is a solution and, by (4.3), also the largest one. □

Similarly, using Corollary 2, we easily obtain

Theorem 28 *The fuzzy relation equation*

$$R\mu Y = T \tag{4.4}$$

has a solution Y iff $H_2(R,T)$ is a solution. If a solution exists, then $H_2(R,T)$ is the largest one.

We leave for the reader as an exercise to verify that also the following two theorems hold.

Theorem 29 *A system of fuzzy relation equations*

$$X\mu S_i = T_i,\ i = 1, \cdots, n \tag{4.5}$$

has a solution X iff $C = \cap_{i=1}^{n} H_1(S_i,T_i)$ is a solution. If a solution exists, then C is the largest one.

Theorem 30 *A system of fuzzy relation equations*

$$R_i \mu Y = T_i,\ i = 1, \cdots, n \tag{4.6}$$

has a solution Y iff $D = \cap_{i=1}^{n} H_2(R_i, T_i)$ is a solution. If a solution exists, then D is the largest one.

Theorem 31 *The fuzzy relation equation* (4.2) *has a solution X for each $T \lessapprox U \times W$ iff $H_1(S, I)\mu S = I$.*

Proof. If (4.2) has a solution X for each T, it has it in particular for $T = I$ and hence

$$H_1(S, I)\mu S = I. \tag{4.7}$$

On the other hand if (4.7) holds, we have for $X = T\mu H_1(S, I)$,

$$X\mu S = T\mu(H_1(S, I)\mu S) = T,$$

and the theorem has been proved. □

Similarly, we have

Theorem 32 *The fuzzy relation equation* (4.4) *has a solution Y for each $T \lessapprox U \times W$ iff $R\mu H_2(R, I) = I$.*

Let (4.2) have a solution and $S\mu H_2(S, I) = I$. If $X\mu S = X'\mu S = T$, then $X\mu(S\mu H_2(S, I)) = X'\mu(S\mu H_2(S, I))$, hence $X = X'$. We thus have

Theorem 33 *If the equation* (4.2) *has a solution and $S\mu H_2(S, I) = I$, then the solution is unique.*

Similarly we see

Theorem 34 *If the equation* (4.4) *has a solution and $R\mu H_1(R, I) = I$, then the solution is unique.*

Example 1 *Similarity* is such a fuzzy relation $S \lessapprox U \times U$ that, for each $u, v, w \in U$,
(i) $S\langle u, u\rangle = \mathbf{1}$ ('everything is similar to itself'),
(ii) $S\langle u, v\rangle = S\langle v, u\rangle$ ('similarity is symmetric'),
(iii) $\mu(S\langle u, v\rangle, S\langle v, w\rangle) \leq S\langle u, w\rangle$ ('similarity is weakly transitive').
It is easy to see that similarity is a generalization of classical equivalence relation. If, for example, μ stands for the Łukasiewicz product or Brouwerian product, then the following fuzzy relation S defined on a set $\{a, b, c\}$ is a similarity relation.

$S\langle a, b\rangle =$

	a	b	c
a	1.00	0.80	0.72
b	0.80	1.00	0.90
c	0.72	0.90	1.00

Example 2 Let R and T be two fuzzy relations defined by the following tables

$R\langle u,v\rangle =$

	v_1	v_2	v_3
u_1	0.8	0.7	0.9
u_2	0.8	1.0	0.6
u_3	0.9	1.0	0.9

$T\langle u,w\rangle =$

	w_1	w_2	w_3
u_1	0.8	0.8	0.9
u_2	0.8	1.0	0.9
u_3	0.9	1.0	0.9

We are looking for solutions of the fuzzy relation equation $R \odot Y = T$, where the composition is the Łukasiewicz product composition. By Theorem 28, if a solution exists, then $H_2(R,T)$ is a solution. The fuzzy relation $H_2(R,T)$ is defined by

$$H_2(R,T)\langle v_i,w_j\rangle = \min_{u\in U}\{[1 - R\langle u,v_i\rangle + T\langle u,w_j\rangle] \wedge 1\},\ \ i,j = 1,2,3.$$

By direct computation we obtain

$H_2(R,T)\langle v,w\rangle =$

	w_1	w_2	w_3
v_1	1.0	1.0	1
v_2	0.8	1.0	0.9
v_3	0.9	0.9	1.0

By definition, for $i,j = 1,2,3$,

$$R \odot H_2(R,T)\langle u_i,w_j\rangle = \max_{v\in V}\{[R\langle u_i,v\rangle + H_2(R,T)\langle v,w_j\rangle - 1] \vee 0\}.$$

It is now a routine work to check that $R \odot H_2(R,T)\langle u_i,w_j\rangle = T\langle u_i,w_j\rangle$ for $i,j = 1,2,3$. It should be noticed, however, that $H_2(R,T)$ is not the only solution. Indeed, e.g.

$S\langle v,w\rangle =$

	w_1	w_2	w_3
v_1	1.0	0.8	0.6
v_2	0.8	1.0	0.9
v_3	0.6	0.9	1.0

is another solution.

Example 3 Now we ask if the fuzzy relation equation in Example 2, where R is fixed, has a solution for each T. By Theorem 32, this happens only if $R \odot H_2(R,I) = I$, where the identity relation I is the following

$I\langle v,v\rangle =$

	v_1	v_2	v_3
v_1	1	0	0
v_2	0	1	0
v_3	0	0	1

First we calculate $H_2(R,I)$ and obtain

$H_2(R,I)\langle u,v\rangle =$

	v_1	v_2	v_3
u_1	0.1	0.1	0.2
u_2	0.0	0.2	0.0
u_3	0.1	0.1	0.1

Since e.g. $R \odot H_2(R,I)\langle v_1, v_1\rangle = 0 \neq 1$ we conclude that there exists such fuzzy relations T that $R \odot Y = T$ has no solution Y.

Example 4 Repeat Example 2 by considering now the sub-min composition between fuzzy relations, i.e. μ is the Brouwer product. Then, by definition,

$$H_2(R,T)\langle v_i, w_j\rangle = \min_{u\in U}\{R\langle u, v_i\rangle \to T\langle u, w_j\rangle\},\ i,j = 1,2,3,$$

where

$$R\langle u, v_i\rangle \to T\langle u, w_j\rangle = \begin{cases} 1 & \text{for } R\langle u, v_i\rangle \leq T\langle u, w_j\rangle \\ T\langle u, w_j\rangle & \text{otherwise.} \end{cases}$$

Then

$H_2(R,T)\langle v,w\rangle =$

	w_1	w_2	w_3
v_1	1.0	1.0	1.0
v_2	0.8	1.0	0.9
v_3	0.8	0.8	1.0

and the fuzzy relation $R \odot H_2(R,T)$ is given, for $i,j = 1,2,3$, by

$$R \odot H_2(R,T)\langle u_i, w_j\rangle = \max_{v\in V}\{R\langle u_i, v\rangle \wedge H_2(R,T)\langle v, w_j\rangle\}.$$

We realize that $R \odot H_2(R,T) = T$. Therefore the fuzzy relation equation under consideration has at least one solution and, by Theorem 28, $H_2(R,T)$ is the largest one.

4.2 On Fuzzy Similarity Relations

For now on, we assume all the time that L is a complete BL-algebra and take a better look at similarity relation that we introduced in Example 5.1.1. Depending on the choice of the operation $\odot$, this relation is called by different names in literature, e.g. *fuzzy equivalence relation, indistinguishability operator, fuzzy equality* or *proximity relation.* An *L-fuzzy subset* (or simply *fuzzy set*) of a set X is a mapping $\mu : X \mapsto L$, where μ is called *membership function.* The collection of all L-fuzzy subsets of a set X is denoted by L^X. Since L is ordered we can define an order relation $\leq$ on the collection L^X in the obvious way.

Example 1 In everyday language we are used to talk about young, middle aged and old people. The borders of these subsets of human beings are, however, vague and unclear. We may illustrate these fuzzy subsets of humans in the following way:

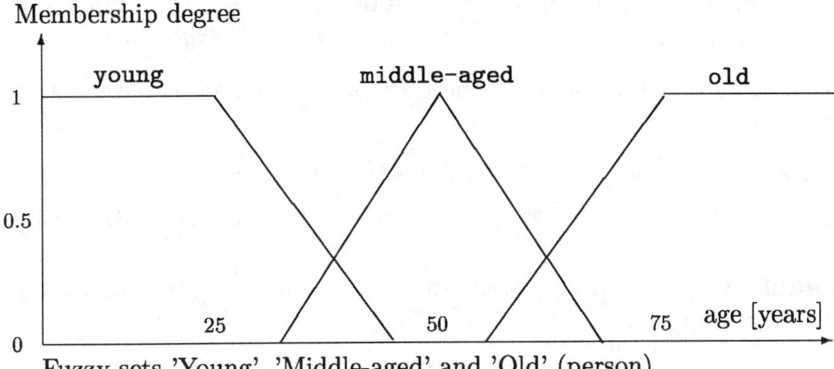

Fuzzy sets 'Young', 'Middle-aged' and 'Old' (person).

$$
\begin{array}{lll}
\mu_{\text{young}}(age = 20) = 1 & \mu_{\texttt{mid.}}(age = 20) = 0 & \mu_{\text{old}}(age = 20) = 0 \\
\mu_{\text{young}}(age = 38) = 0.4 & \mu_{\texttt{mid.}}(age = 38) = 0.4 & \mu_{\text{old}}(age = 38) = 0 \\
\mu_{\text{young}}(age = 63) = 0 & \mu_{\texttt{mid.}}(age = 63) = 0.4 & \mu_{\text{old}}(age = 63) = 0.4 \\
\mu_{\text{young}}(age = 80) = 0 & \mu_{\texttt{mid.}}(age = 80) = 0 & \mu_{\text{old}}(age = 80) = 1
\end{array}
$$

Example 2 For a human being, to be called *tall* or *fat* seems to be a cultural and time-dependent phenomena. Besides, a person needs to be both tall and fat to be called *massive*. By fuzzy sets,

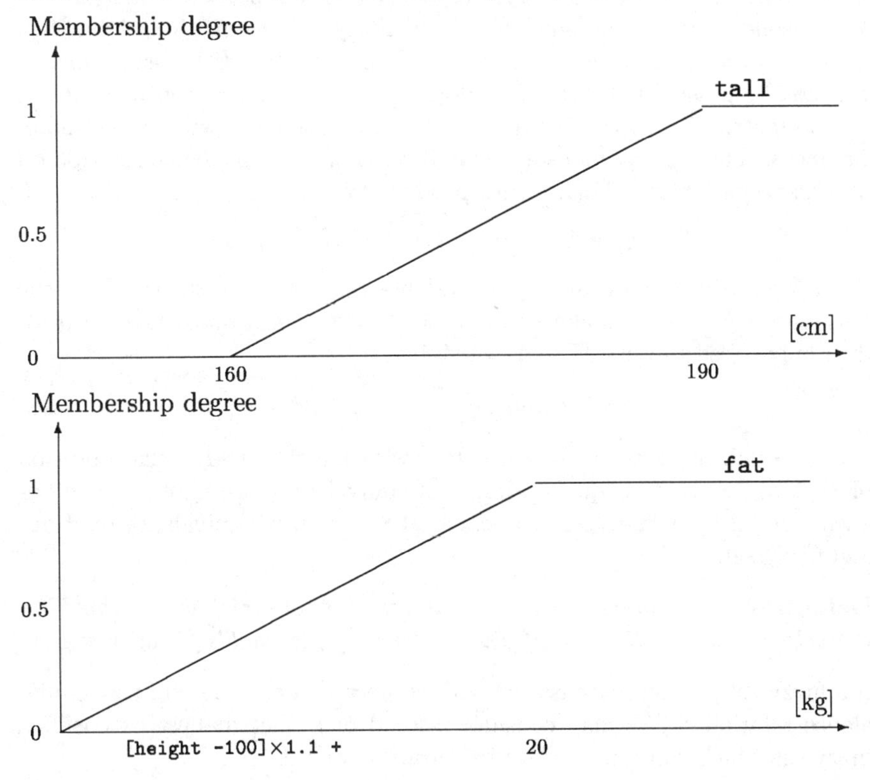

From these more or less subjective definition we can see, for example, that

$$\mu_{\mathtt{tall}}(height = 187cm) = 0.9 \quad \mu_{\mathtt{fat}}(weigh = 103kg) = 0.4.$$

If we assume that tallness and fatness are both factors of massiveness, we may count

$$\mu_{\mathtt{massive}}(h = 187, w = 103) = \mu_{\mathtt{tall}}(height = 187) \odot \mu_{\mathtt{fat}}(weigh = 103).$$

E.g. in Łukasiewicz structure $\mu_{\mathtt{massive}}(height = 187, weigh = 103) = 0.3$.

Example 3 If $\sim$ an ordinary equivalence relation on X, then its characteristic function

$$S_R(x,y) = \begin{cases} 1 & \text{if } x \sim y \\ 0 & \text{otherwise,} \end{cases}$$

is a fuzzy equivalence relation with respect to any choice of the operation $\odot$. Especially the characteristic function of the crisp equality on X is a fuzzy equivalence relation.

From general topology we know that *pseudo-metric* on a set X is a function d on the Cartesian product $X \times X$ to the non-negative real numbers such that, for all points x, y and z of X, (a) $d(x,y) = d(y,x)$, (b) $d(x,z) \leq d(x,y) + d(y,z)$ and (c) $d(x,x) = 0$. Let $\odot$ be the Łukasiewicz product. It is an exercise to show that S is a fuzzy equivalence relation on X with respect to $\odot$ if, and only if $1 - S$ is a pseudo-metric on X. Thus pseudo-metrics bounded by 1 and fuzzy equivalence relations with respect to the Łukasiewicz product are dual concepts. If a pseudo-metric d is not bounded by 1, we can enforce this property by considering the pseudo-metric $\hat{d}(x,y) = \min\{d(x,y), 1\}$, which coincides with d for 'small' distances. Thus any pseudo-metric induces a fuzzy equivalence relation on X with respect to the Łukasiewicz product by

$$S(x,y) = 1 - \hat{d}(x,y) = 1 - \min\{d(x,y), 1\}.$$

When we interpret an ordinary equivalence relation $\sim$ on the set X in the sense that equivalent elements cannot be distinguished or may be identified, then only those subsets $M \subseteq X$ satisfying

$$\text{if } x \in M \text{ and } x \sim y \text{ then } y \in M \tag{4.8}$$

'behave well' with respect to $\sim$. In other words, a subset M fulfils condition (4.8) if, and only if it equals a union of equivalence classes of $\sim$. The following definition generalizes the axiom (4.8) for fuzzy equivalence relations and fuzzy sets.

Definition 32 *A fuzzy set $\mu \in L^X$ is called* extensional *w.r.t. the fuzzy equivalence relation S on X iff $\mu(x) \odot S(x,y) \leq \mu(y)$ holds for all $x, y \in X$.*

If a fuzzy set μ is not extensional with respect to a considered fuzzy equivalence relation S, we may consider instead of μ the smallest extensional fuzzy set which contains μ. More accurately, we set

Definition 33 *Let S be a fuzzy equivalence relation on X and let $\mu \in L^X$. The fuzzy set*

$$\widehat{\mu} = \bigwedge\{\nu \mid \mu \leq \nu \text{ and } \nu \text{ is extensional w.r.t. } S\}$$

is called the extensional hull *of μ w.r.t. S.*

Proposition 101 *Let S be a fuzzy equivalence relation on X and let $\mu \in L^X$. Then*

$$\widehat{\mu}(x) = \bigvee\{\mu(y) \odot S(x,y) \mid y \in X\}, \tag{4.9}$$

$$\widehat{\mu} \text{ is extensional w.r.t. } S, \tag{4.10}$$

$$\widehat{\widehat{\mu}} = \widehat{\mu}. \tag{4.11}$$

Proof. Let us abbreviate the right hand side of (4.9) by $\widetilde{\mu}$. $\widetilde{\mu}$ is extensional w.r.t. S because

$$\begin{aligned}
\widetilde{\mu}(x) \odot S(x,y) &= S(x,y) \odot \bigvee\{\mu(z) \odot S(x,z) \mid z \in X\} \\
&= \bigvee\{\mu(z) \odot S(y,x) \odot S(x,z) \mid z \in X\} \\
&\leq \bigvee\{\mu(z) \odot S(y,z) \mid z \in X\} \\
&= \widetilde{\mu}(y).
\end{aligned}$$

We also have that

$$\mu(x) = \mu(x) \odot S(x,x) \leq \bigvee\{\mu(y) \odot S(x,y) \mid y \in X\} = \widetilde{\mu}(x),$$

which implies $\widehat{\mu} \leq \widetilde{\mu}$. In order to prove $\widetilde{\mu} \leq \widehat{\mu}$, let ν be an extensional fuzzy set of X w.r.t. S such that $\mu \leq \nu$. Since

$$\mu(y) \odot S(x,y) \leq \nu(y) \odot S(x,y) \leq \nu(y)$$

holds for any $y \in X$, we have also $\widetilde{\mu} \leq \widehat{\mu}$. This proves (4.9). (4.10) and (4.11) now follow by (4.9) and the definition of $\widehat{\mu}$.□

Proposition 101 states that $\widehat{\mu}$ can be interpreted as the union of all elements that are equivalent w.r.t. S to at least one of the elements of μ.

Theorem 35 *Let S be a fuzzy equivalence relation on X and let $\mu \in L^X$. Then μ is extensional w.r.t. S iff for all $x, y \in X$*

$$S(x,y) \leq \mu(x) \leftrightarrow \mu(y). \tag{4.12}$$

Proof. Let μ be extensional w.r.t. S. Then $S(x,y) \leq \mu(x) \leftrightarrow \mu(y)$ iff $S(x,y) \leq [\mu(x) \to \mu(y)] \wedge [\mu(y) \to \mu(x)]$ iff $S(x,y) \leq [\mu(x) \to \mu(y)]$ and $S(x,y) \leq [\mu(y) \to \mu(x)]$ iff $\mu(x) \odot S(x,y) \leq \mu(y)$ and $\mu(y) \odot S(y,x) \leq \mu(x)$, which holds true. The converse is now trivial.□

Example 4 We can define the extensional hull $\widehat{M}$ of a crisp subset M of X w.r.t. any fuzzy equivalence relation S as the extensional hull w.r.t. S of its characteristic function, i.e.

$$\widehat{M} = \begin{cases} 1 & \text{if } x \in M \\ 0 & \text{otherwise.} \end{cases}$$

Let X be the set of real numbers, and let $\odot$ be the Łukasiewicz product. Consider the fuzzy equivalence relation induced by the usual metric on real numbers, i.e. $S(x, y) = 1 - \min\{|x - y|, 1\}$. Then the extensional hull w.r.t. S of the one-element set $\{x_0\}$ is the *triangular fuzzy set*

$$\widehat{\{x_0\}} = 1 - \min\{|x_0 - x|, 1\},$$

and the extensional hull w.r.t. S of an interval $[a, b]$ is the *trapezoidal fuzzy set*

$$\widehat{[a,b]}(x) = \begin{cases} 1 & \text{if } a \leq x \leq b \\ \max\{1 - a + x, 0\} & \text{if } x \leq a \\ \max\{1 - x + b, 0\} & \text{if } b \leq x. \end{cases}$$

By using the scaling function of the usual metric on the real numbers triangular and trapezoidal fuzzy sets with other slopes that 1 can be obtained as extensional hull w.r.t. S of single elements and intervals, respectively.

We have now seen how a crisp set induces a fuzzy set as its extensional hull w.r.t. to a fuzzy equivalence relation. Thus, assuming the indistinguishability modeled by a fuzzy equivalence relation as a basic concept, fuzzy sets can be viewed as induced concepts, i.e. we obtain membership degrees starting from (fuzzy) indistinguishability. Now we take a closer look at the connection between fuzzy sets and the corresponding indistinguishability. As a main result, we show how the indistinguishability inherent to a given collection of fuzzy set can be derived.

Theorem 36 *Let $\mathcal{F} \subseteq L^X$ be a collection of fuzzy sets. Then*

$$S_{\mathcal{F}}(x, y) = \bigwedge_{\mu \in \mathcal{F}} [\mu(x) \leftrightarrow \mu(y)] \tag{4.13}$$

is the coarsest (greatest) fuzzy equivalence relation on X such that all fuzzy sets in $\mathcal{F}$ are extensional w.r.t. $S_{\mathcal{F}}$.

Proof. It is obvious that $S_{\mathcal{F}}$ is reflexive and symmetric. The transitivity of $S_{\mathcal{F}}$ follows from

$$
\begin{aligned}
S_{\mathcal{F}}(x,y) \odot S_{\mathcal{F}}(y,z) &= \{\bigwedge_{\mu\in\mathcal{F}} [\mu(x) \leftrightarrow \mu(y)]\} \odot \{\bigwedge_{\nu\in\mathcal{F}} [\nu(y) \leftrightarrow \nu(z)]\} \\
&\leq \bigwedge_{\mu,\nu\in\mathcal{F}} [\mu(x) \leftrightarrow \mu(y)] \odot [\nu(y) \leftrightarrow \nu(z)] \\
&\leq \bigwedge_{\mu\in\mathcal{F}} [\mu(x) \leftrightarrow \mu(y)] \odot [\mu(y) \leftrightarrow \mu(z)] \\
&\leq \bigwedge_{\mu\in\mathcal{F}} [\mu(x) \leftrightarrow \mu(y)] \\
&= S_{\mathcal{F}}(x,z),
\end{aligned}
$$

where we made use of the facts proved in Exercise 1.3.20 and (1.59). The extensionality of the fuzzy sets in $\mathcal{F}$ follows directly from Theorem 35 and the definition of $S_{\mathcal{F}}$. Finally, we have to show that $S_{\mathcal{F}}$ is the coarsest fuzzy equivalence relation making all fuzzy sets in $\mathcal{F}$ extensional. Let S be a fuzzy equivalence relation such that all fuzzy sets in $\mathcal{F}$ are extensional w.r.t. S. By Theorem 35, $S(x,y) \leq \mu(x) \leftrightarrow \mu(y)$ holds for all $\mu \in \mathcal{F}$ which implies $S(x,y) \leq S_{\mathcal{F}}(x,y)$.□

The formula (4.13) can be interpreted in the following way. Two elements 'cannot be distinguished by a (fuzzy) set' if they are either both elements of the same set or its complement, but not one in the set and another one in the complement. Thus, the value $\mu(x) \leftrightarrow \mu(y)$ represents the degree to which the elements x and y cannot be distinguished by the fuzzy set μ. Therefore $S_{\mathcal{F}}(x,y)$ is the degree to which x and y are similar with respect to the collection $\mathcal{F}$ of fuzzy sets.

Example 5 The socio-economic status of a person depends on, say, three factors: income, personal property and status in working life, which are typical fuzzy sets. Assume we want to compare three persons' X, Y, Z socio-economic status having available the following information.

Person	Income [$/month]	Property [$]	Occupation
X	2300	8000	researcher
Y	1500	1.000.000	teacher
Z	3000	600.000	journalist

All these three persons have quite high status in working life, thought the researcher has the highest one of them and the teacher has the lowest one. With respect to monthly income, Z has relative high income, Y does not have high income and X has income something in between these two. Y can be called rich while X is not rich at all and Z is more or less rich. By fuzzy sets,

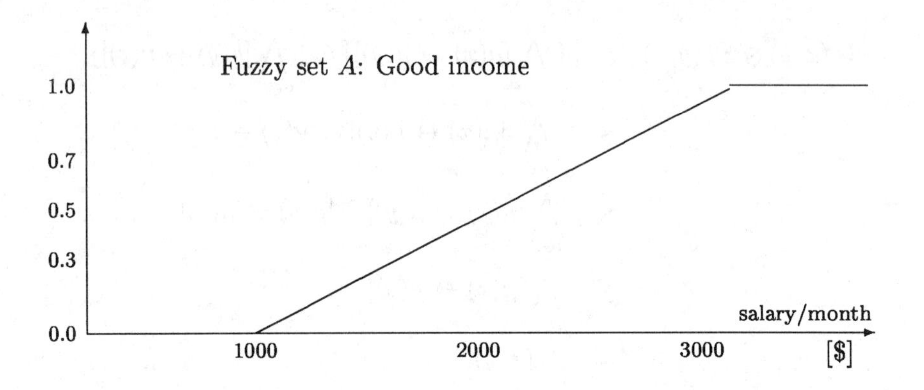

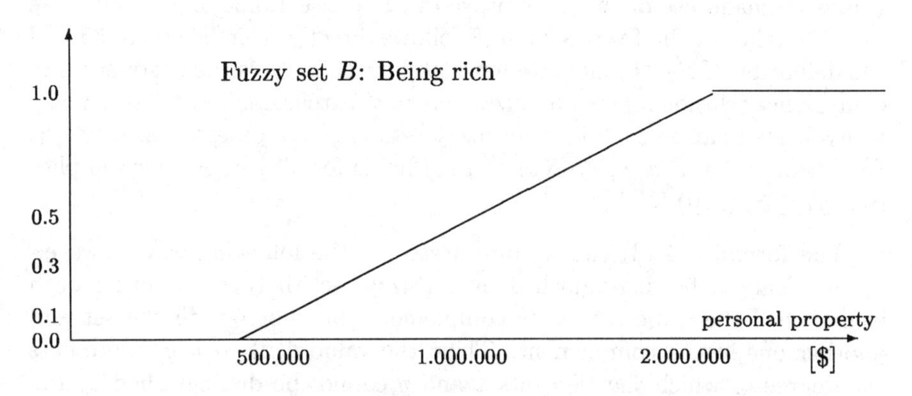

The fuzzy set C 'High status in working life' can be defined as a discrete fuzzy set. Then the membership degrees of the persons X, Y and Z in these fuzzy sets are the following

$$\begin{array}{lll} \mu_A(X) = 0.5 & \mu_B(X) = 0 & \mu_C(X) = 0.9 \\ \mu_A(Y) = 0.3 & \mu_B(Y) = 0.3 & \mu_C(Y) = 0.7 \\ \mu_A(Z) = 0.7 & \mu_B(Z) = 0.1 & \mu_C(Z) = 0.8 \end{array}$$

Now a socio-economic similarity relation S is counted, according to (4.13), by

$$S(a,b) = \min\{\mu_k(a) \leftrightarrow \mu_k(b) \mid k \in \{A,B,C\}\}, \text{ where } a, b \in \{X,Y,Z\}.$$

The result depends, of course, on the choice of the BL-algebra L. Taking, for example, L a generalized Łukasiewicz structure, $p = 2$, we obtain

$S\langle a,b\rangle$	X	Y	Z
X	1.00	0.82	0.87
Y	0.82	1.00	0.77
Z	0.87	0.77	1.00

Here one similarity relation is generated by three fuzzy sets. In the next section we will see that, in certain cases, we may combine the three fuzzy similarities generated by each fuzzy set to obtain a fourth similarity relation. This method will offer us even better models.

Interpreting the elements a of the BL-algebra L as constant fuzzy set, we have the following

Theorem 37 *Let S be a fuzzy equivalence relation on the set X. Let $\mathcal{A}_S \subseteq L^X$ denote the collection of fuzzy sets that are extensional w.r.t. S. For all $\mu \in \mathcal{A}_S$, $\mathcal{B} \subseteq \mathcal{A}_S$, $a \in L$, the following statements are valid*

$$\bigvee \mathcal{B} \in \mathcal{A}_S, \tag{4.14}$$

$$\bigwedge \mathcal{B} \in \mathcal{A}_S, \tag{4.15}$$

$$(a \odot \mu) \in \mathcal{A}_S, \tag{4.16}$$

$$(\mu \to a) \in \mathcal{A}_S, \tag{4.17}$$

$$(a \to \mu) \in \mathcal{A}_S, \tag{4.18}$$

Proof. (4.14) follows directly from (1.28), (4.15) is implied by Exercise 1.3.20 and (4.16) is obvious. For (4.17), we have to prove that

$$S(x,y) \odot [\mu(x) \to a] \leq \mu(y) \to a$$

for any $\mu \in \mathcal{A}_S$. In accordance to the adjunction property, this is equivalent to prove that

$$\mu(y) \odot S(x,y) \odot [\mu(x) \to a] \leq a,$$

which is the case as, by the extensionality of μ, $\mu(y) \odot S(x,y) \leq \mu(x)$. Finally, by adjunction, it is sufficient to prove that

$$S(x,y) \odot a \odot [a \to \mu(x)] \leq \mu(y),$$

for (4.18), which holds true because of the extensionality of μ and the fact that $a \odot [a \to \mu(x)] \leq \mu(x)$.□

Theorem 38 *Let $\mathcal{A} \subseteq L^X$ be such a collection of fuzzy sets that properties* (4.14) - (4.18) *hold. Then there exists a fuzzy equivalence relation S on X for which the corresponding collection of extensional fuzzy sets is exactly $\mathcal{A}$, i.e. $\mathcal{A}_S = \mathcal{A}$. Furthermore, S is uniquely determined by equation* (4.13) *with $\mathcal{F} = \mathcal{A}$.*

Proof. First we prove that $\mathcal{A}_S = \mathcal{A}$ for the fuzzy equivalence relation

$$S(x,y) = \textstyle\bigwedge_{\mu \in \mathcal{A}} [\mu(x) \leftrightarrow \mu(y)].$$

By the definition of S and in accordance to Theorem 35, all fuzzy sets in $\mathcal{A}$ are extensional w.r.t. S, i.e. $\mathcal{A} \subseteq \mathcal{A}_S$. In order to show the other inclusion, let $\mu \in \mathcal{A}_S$. For $z \in X$, define the fuzzy set μ_z by setting for any $x \in X$:

$$\mu_z = \mu(z) \odot S(x, z).$$

Then

$$\mu_z = \bigwedge\nolimits_{\nu \in \mathcal{A}} [\mu(z) \odot (\nu(x) \leftrightarrow \nu(z))].$$

Due to the closure properties (4.15) - (4.18) of $\mathcal{A}$, we have that $\mu_z \in \mathcal{A}$. Moreover, since μ is extensional w.r.t. S we obtain $\mu_z(x) \leq \mu(x)$. This, together with $\mu_z(z) = \mu(z) \odot S(z, z) = \mu(z)$ implies $\mu = \bigvee_{z \in X} \mu_z$, which again is an element of $\mathcal{A}$ by the property (4.14). Thus we have proved the other inclusion, too, concluding $\mathcal{A}_S = \mathcal{A}$. Let us now turn to the uniqueness of S. Let $\widetilde{S}$ be a fuzzy equivalence relation such that $\mathcal{A}_{\widetilde{S}} = \mathcal{A}$. In accordance to Theorem 36, we have $\widetilde{S} \leq S$. To show that $S \leq \widetilde{S}$ holds, too, let $x, y \in X$. Define the fuzzy set $\nu_y(z) = \widetilde{S}(z, y)$, which is obviously extensional w.r.t. $\widetilde{S}$, therefore it is an element of $\mathcal{A}$ and thus, extensional w.r.t. S, too. Furthermore, we have $\nu_y(y) = \mathbf{1}$. We finally conclude that

$$S(x, y) = S(x, y) \odot \mathbf{1} = S(x, y) \odot \nu_y(y) \leq \widetilde{S}(x, y),$$

and the proof is complete.□

Theorems 37 and 38 establish a one-to-one correspondence between the fuzzy equivalence relations on a set X and the collection of all fuzzy sets on X fulfilling the closure properties (4.14) - (4.18). The conditions (4.14) and (4.15) state that extensionality is preserved by arbitrary unions and intersections and the conditions (4.16) and (4.18) can be interpreted as a kind of cutting and lifting conditions, respectively, that maintain extensionality. The condition (4.17) means that extensionality is preserved under a generalized complementation, where $a = \mathbf{0}$ corresponds to the usual complementation. For MV-algebras the conditions (4.14) - (4.18) can be simplified.

Theorem 39 *Let L be a complete MV-algebra and $\mathcal{A} \subseteq L^X$ such a collection of fuzzy sets that*

$$\text{for all } \mathcal{B} \subseteq \mathcal{A}, \bigvee \mathcal{B} \in \mathcal{A}, \tag{4.19}$$

$$\text{for all } a \in L, \mu \in \mathcal{A}, (\mu \to a) \in \mathcal{A}. \tag{4.20}$$

Then $\mathcal{A}$ satisfies (4.14) - (4.18).

Proof. (4.14) is identical with (4.19) and, as in MV-algebras holds $\bigwedge \mathcal{B}$ $= [\bigvee_{\nu \in \mathcal{B}}](\nu \to \mathbf{0}) \to \mathbf{0}$ we have (4.15). Since product in MV-algebras is expressible by means of residuum, i.e. $\mu \odot a = [\mu \to (a \to \mathbf{0})] \to \mathbf{0}$, we have (4.16). (4.17) is just (4.20) and (4.18) holds as in MV-algebras $a \to \mu =$ $[a \odot (\mu \to \mathbf{0})] \to \mathbf{0}]$.□

Theorem 39 shows that when the underlying BL-algebra is a complete MV-algebra, the algebraic characterization of the collection of extensional fuzzy sets simplifies to the closeness with respect to arbitrary unions and to the generalized complementation $\mu \to a$.

In this section we have discussed connections between fuzzy sets and fuzzy equivalence relations. For a given collection of fuzzy sets, the fuzzy equivalence relation (4.13) characterizes the indistinguishability inherent to these fuzzy sets. In the next section we shall see that this indistinguishability cannot be overcome in typical approximate reasoning situations with fuzzy sets.

4.3 Fuzzy Similarity and Approximate Reasoning

The aim of this section is to show that in typical applications of fuzzy reasoning, for example in fuzzy control, fuzzy equivalence relations are of importance since they characterize an indistinguishability that cannot be overcome. In *approximate reasoning* one deals with IF-THEN rules of the form

$$\text{If } x \text{ is } A \text{ then } y \text{ is } B, \tag{4.21}$$

where x and y are variables with domains X and Y, respectively, A and B are so-called *linguistic terms* like `high temperature` or `approximately zero`. These linguistic terms are usually modeled by suitable fuzzy sets, say $\mu_A \in L^X$, $\mu_B \in L^Y$.

Example 1 Assume we have a simple control system for heating a house. Then an obvious fuzzy IF-THEN role would be

IF temperature is `high` THEN heating is `approximately zero`.

and we should have at least the following fuzzy sets

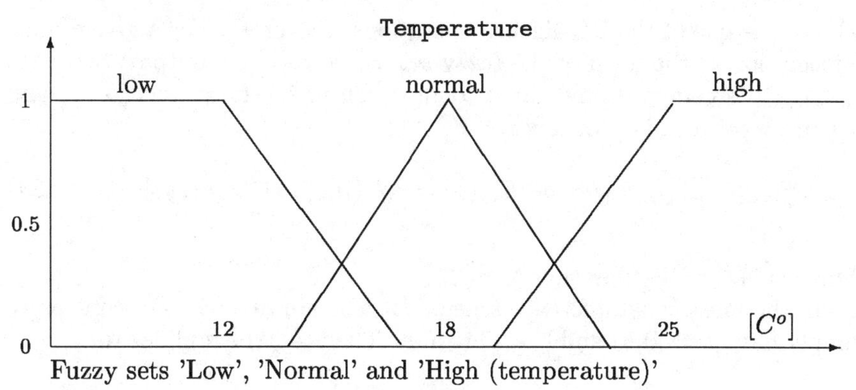

Fuzzy sets 'Low', 'Normal' and 'High (temperature)'

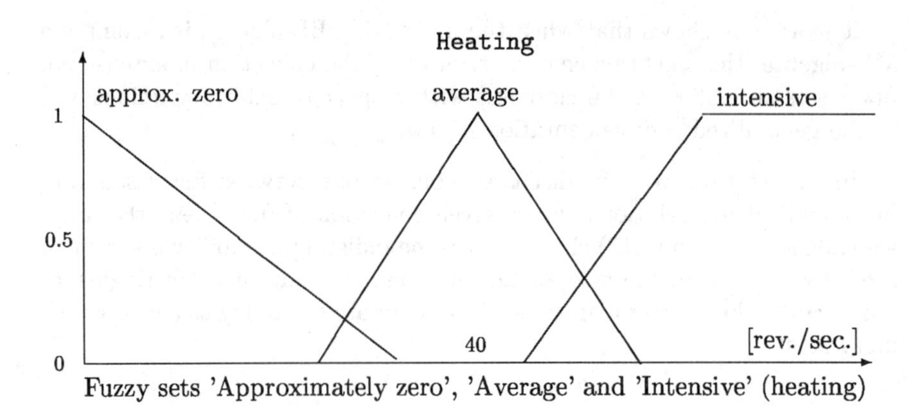

Fuzzy sets 'Approximately zero', 'Average' and 'Intensive' (heating)

Rules of the form (4.21) represent general knowledge about the considered problem. In an actual situation, in addition to this general knowledge, the information

$$x \text{ is } A' \tag{4.22}$$

is given, where A' is represented by a fuzzy set $\mu_{A'} \in L^X$. Of course, as a special case $\mu_{A'}$ can also stand for a crisp value x_0 in which case $\mu_{A'}(x)$ would yield the value **1** for $x = x_0$ and **0** otherwise. Thus, an inference scheme is needed that derives from the fuzzy sets μ_A, μ_B, appearing in the rules, and the fuzzy set $\mu_{A'}$, representing the actual information, a fuzzy set $\nu_{\text{conclusion}}$ that describes the restriction or possible values for the variable y under the given rules and the actual input information (4.22).

Let us first only consider a single rule of the form (4.21). A very convenient possibility to represent such a rule is in the form of a fuzzy relation $R_* \in L^{X \times Y}$. Usually R_* is defined by

$$R_*(x, y) = \mu_A(x) \Box \mu_B(y), \tag{4.23}$$

where $\Box$ is one of the BL-algebra operations $\wedge, \odot$ or $\rightarrow$. For a given input information in the form of the fuzzy set $\mu_{A'} \in L^X$, the 'output' fuzzy set $\nu_{\text{conclusion}}$ is computed by sup-$\sqcap$ composition of the fuzzy relation R and the fuzzy set $\mu_{A'}$, i.e. for any $y \in Y$,

$$\nu_{\text{conclusion}}(y) = (\mu_{A'} \circ_{\sqcap} R_*)(y) = \bigvee_{x \in X} \{\mu_{A'}(x) \sqcap R_*(x, y)\}, \tag{4.24}$$

where $\sqcap$ is the BL-operation $\wedge$ or $\odot$.

In the most usual inference scheme for a single rule in fuzzy control we simply have $L = [0, 1]$ and $\sqcap = \Box = \min$. Then (4.24) simplifies to

$$\nu_{\text{conclusion}}(y) = \bigvee_{x \in X} \min\{\mu_{A'}(x), \mu_A(x), \mu_B(y)\}, \tag{4.25}$$

and if $\mu_{A'}$ is just the characteristic function of a single element set $\{x_0\}$ then (4.25) reads even more simply

$$\nu_{\text{conclusion}}(y) = \min\{\mu_A(x_0), \mu_B(y)\}, \tag{4.26}$$

as in this case

$$R_*(x, y) = \min\{\mu_A(x), \mu_B(y)\} \tag{4.27}$$

Another interesting and useful case is when $\square$ is the Brouwer implication, that is

$$R_*(x, y) = \begin{cases} 1 & \text{if } \mu_A(x) \leq \mu_B(y) \\ \mu_B(y) & \text{otherwise,} \end{cases} \tag{4.28}$$

and $\sqcap$ stands for min.

Notice that it is not reasonable to consider all possible combinations of the choices for $\sqcap$ and $\square$ in (4.24). It will be an exercise for the reader to prove the following

Remark 26 *In case of $\mu_A = \mu_{A'}$ and there is an $x \in X$ such that $\mu_A(x) = 1$ (such a fuzzy set is called* normal*), then*

$$\nu_{\text{conclusion}}(y) = \mu_B(y) \text{ for all } y \in Y, \tag{4.29}$$

if (i) $\square = \rightarrow$ *and* $\sqcap = \odot$ *or* (ii) $\sqcap = \odot$ *and* $\square = \odot$ *or* (iii) $\square = \odot$ *and* $\sqcap = \wedge$ *or* (iv) $\square = \wedge$ *and* $\sqcap = \odot$ *or* (v) $\sqcap = \wedge$ *and* $\square = \wedge$.

For the missing combination $\square = \rightarrow$ and $\sqcap = \wedge$ (4.29) does in general not hold. Indeed, consider the Łukasiewicz structure, $X = Y = \{a, b\}$ and $\mu_A(a) = 1$, $\mu_A(b) = 0.9$, $\mu_B(a) = 1$, $\mu_B(b) = 0.8$. Then $\nu_{\text{conclusion}}(b) = 0.9 \neq \mu_B(b) = 0.8$.

We are now prepared to formulate two interesting and important theorems about applying IF-THEN rules and indistinguishability inherent to the corresponding fuzzy sets.

Theorem 40 *Let $\mu_{A'}, \mu_A \in L^X$, $\mu_B \in L^Y$ and S a fuzzy equivalence relation on X such that μ_A is extensional w.r.t. S. Furthermore, let R_* be defined by equation* (4.23)*. Then, for the combinations* (i) $\square = \rightarrow$ *and* $\sqcap = \odot$, (ii) $\square = \odot$ *and* $\sqcap = \odot$, (iii) $\square = \wedge$ *and* $\sqcap = \odot$ *holds* $\mu_{A'} \circ_\sqcap R_* = \widehat{\mu_{A'}} \circ_\sqcap R_*$.

Proof. $\mu_{A'} \leq \widehat{\mu_{A'}}$, together with the isotonicity of $\sqcap$, implies that $\mu_{A'} \circ_\sqcap R_* \leq \widehat{\mu_{A'}} \circ_\sqcap R_*$ for any choice of $\sqcap$ and $\square$. To establish the other direction, we proceed by cases. Denote $A = \widehat{\mu_{A'}} \circ_\sqcap R_*(y)$, where $y \in Y$.

Case (i) $\square = \rightarrow$ and $\sqcap = \odot$. Then

$$
\begin{aligned}
A &= \bigvee_{x \in X} \{\widehat{\mu_{A'}}(x) \odot [\mu_A(x) \rightarrow \mu_B(y)]\} \\
&= \bigvee_{x,x' \in X} \{\mu_{A'}(x') \odot S(x,x') \odot [\mu_A(x) \rightarrow \mu_B(y)]\} \\
&\leq \bigvee_{x,x' \in X} \{\mu_{A'}(x') \odot S(x,x') \odot [(S(x,x') \odot \mu_A(x')) \rightarrow \mu_B(y)]\} \\
&= \bigvee_{x,x' \in X} \{\mu_{A'}(x') \odot S(x,x') \odot [S(x,x') \rightarrow (\mu_A(x') \rightarrow \mu_B(y))]\} \\
&= \bigvee_{x,x' \in X} \{\mu_{A'}(x') \odot [S(x,x') \wedge (\mu_A(x') \rightarrow \mu_B(y))]\} \\
&\leq \bigvee_{x,x' \in X} \{\mu_{A'}(x') \odot [\mu_A(x') \rightarrow \mu_B(y)]\} \\
&= (\mu_{A'} \circ_{\sqcap} R_*)(y).
\end{aligned}
$$

In the second line we used equation (1.28), in the third the extensionality of μ_A and the isotonicity of the residuum on the second variable, in the fourth (1.47), in the fifth (1.70), and in the sixth the isotonicity of $\odot$.

Case (ii) $\square = \sqcap = \odot$. In accordance to (1.28) and the extensionality of μ_A, we have that

$$
\begin{aligned}
A &= \bigvee_{x,x' \in X} [\mu_{A'}(x') \odot S(x,x') \odot \mu_A(x) \odot \mu_B(y)] \\
&\leq \bigvee_{x' \in X} [\mu_{A'}(x') \odot \mu_A(x') \odot \mu_B(y)] \\
&= (\mu_{A'} \circ_{\sqcap} R_*)(y).
\end{aligned}
$$

Case (iii) $\square = \wedge$ and $\sqcap = \odot$.

$$
\begin{aligned}
A &= \bigvee_{x,x' \in X} \{\mu_{A'}(x') \odot S(x,x') \odot [\mu_A(x) \wedge \mu_B(y)]\} \\
&\leq \bigvee_{x,x' \in X} \{\mu_{A'}(x') \odot \{[S(x,x') \odot \mu_A(x)] \wedge [S(x,x') \odot \mu_B(y)]\}\} \\
&\leq \bigvee_{x' \in X} \{\mu_{A'}(x') \odot [\mu_A(x') \wedge \odot \mu_B(y)]\} \\
&= (\mu_{A'} \circ_{\sqcap} R_*)(y).
\end{aligned}
$$

In the second line we applied Exercise 1.3.20.$\square$

When we interpret Theorem 40 in the sense that the fuzzy sets μ_A and μ_B represent linguistic terms A and B, respectively, of an IF-THEN rule of the form (4.21), then it states that for the mentioned combinations of operations for a given input $\mu_{A'}$ the output $\nu_{\text{conclusion}}$ inferred by the rule does not change when we replace $\mu_{A'}$ by its extensional hull. This means that the indistinguishability inherent to the fuzzy set μ_A cannot be avoided, even if the input fuzzy set $\mu_{A'}$ stands for a crisp value. Note that if $\mu_{A'}$ is a crisp set, the choice of the operation $\sqcap$ has no influence in equation (4.24).

From Theorem 40 we derive the answer to the question of fuzzy inputs. A fuzzified input does not change the output of a rule as long as the fuzzy set obtained by the fuzzification is contained in the extensional of the original crisp input value. From Theorem 40 we also learn that it does not make sense to measure more exactly than indistinguishability admits. The cases covered by Theorem 40 include the most common and useful formalizations of IF-THEN rules. It will be a non-trivial exercise to prove that Theorem 40 does not in general hold for the remaining combinations $\square = \rightarrow$, $\sqcap = \wedge$ or $\square = \odot$, $\sqcap = \wedge$ or $\square = \sqcap = \wedge$.

Theorem 40 shows that the indistinguishability induced by the fuzzy set representing the linguistic term in the premise of the rule cannot be overcome. The same holds for the output of the rule, i.e. the output can never be more precise than the indistinguishability induced by the fuzzy set modeling the linguistic term in the conclusion of the rule, as we shall now see.

Theorem 41 *Let $\mu_{A'}, \mu_A \in L^X$, $\mu_B \in L^Y$ and S a fuzzy equivalence relation on Y such that μ_B is extensional w.r.t. S. Furthermore, let R_* be defined by equation* (4.23). *Then, for the combinations* (i) $\square = \rightarrow$ *and* $\sqcap = \odot$, (ii) $\square = \sqcap = \odot$, (iii) $\square = \wedge$ *and* $\sqcap = \odot$, *the fuzzy set* $(\mu_{A'} \circ_\sqcap R_*)$ *is extensional w.r.t. S.*

Proof. Let $A = (\mu_{A'} \circ_\sqcap R_*)(y) \odot S(y, y')$, where $y, y' \in Y$. For the combination (i) $\square = \rightarrow$ and $\sqcap = \odot$ we reason, by Exercise 1.3.21 and the extensionality of μ_B, that

$$\begin{aligned} A &= \bigvee_{x \in X} \{\mu_{A'}(x) \odot [\mu_A(x) \rightarrow \mu_B(y)] \odot S(y, y')\} \\ &\leq \bigvee_{x \in X} \{\mu_{A'}(x) \odot [\mu_A(x) \rightarrow (\mu_B(y) \odot S(y, y'))]\} \\ &\leq \bigvee_{x \in X} \{\mu_{A'}(x) \odot [\mu_A(x) \rightarrow (\mu_B(y')]\} \\ &= (\mu_{A'} \circ_\sqcap R_*)(y'). \end{aligned}$$

For the combination (ii) $\square = \sqcap = \odot$ we have

$$\begin{aligned} A &= \bigvee_{x \in X} \{\mu_{A'}(x) \odot \mu_A(x) \odot \mu_B(y) \odot S(y,y')\} \\ &\leq \bigvee_{x \in X} \{\mu_{A'}(x) \odot \mu_A(x) \odot \mu_B(y')\} \\ &= (\mu_{A'} \circ_\sqcap R_*)(y'). \end{aligned}$$

The claim holds also for the combination (iii) $\Box = \wedge, \sqcap = \odot$, as by Exercise 1.3.20,

$$\begin{aligned} A &= \bigvee_{x \in X} \{\mu_{A'}(x) \odot [\mu_A(x) \wedge \mu_B(y)] \odot S(y,y')\} \\ &\leq \bigvee_{x \in X} \{\mu_{A'}(x) \odot \{[\mu_A(x) \odot S(y,y')] \wedge [\mu_B(y) \odot S(y,y')]\}\} \\ &\leq \bigvee_{x \in X} \{\mu_{A'}(x) \odot \mu_A(x) \odot \mu_B(y')\} \\ &= (\mu_{A'} \circ_\sqcap R_*)(y'). \end{aligned}$$

The proof is complete.□

In Theorems 40 and 41 we have considered only a single IF-THEN rule of the form (4.21). These results can be extended, however, to a more realistic case of a collection of fuzzy IF-THEN rules. Consider the set of rules

$$\text{if } x \text{ is } A_i \text{ then } y \text{ is } B_i \text{ and } i \in \Gamma,$$

where the linguistic terms A_i and B_i are modeled by fuzzy sets $\mu_{A_i} \in L^X$ and $\mu_{B_i} \in L^Y$. Given an 'input fuzzy set' $\mu \in L^X$, the output of this set of rules is computed in the following way: as the first step, a combination of the BL-operations $\Box$ and $\sqcap$ is chosen. Let us assume that one of the three combinations in Theorems 40 and 41 is considered. Then, for each single rule, the corresponding fuzzy relation $R^i_*(x,y) = \mu_{A_i}(x) \Box \mu_{B_i}(y)$ as well as, on this basis, the corresponding output fuzzy relation $\nu_{\text{conclusion}} = \mu \circ_\sqcap R^i_*$ is computed. Finally, these outputs are aggregated by taking

$$\text{either } \bigwedge_{i \in \Gamma} (\mu \circ_\sqcap R^i_*) \tag{4.30}$$

$$\text{or } \bigvee_{i \in \Gamma} (\mu \circ_\sqcap R^i_*). \tag{4.31}$$

Generalizing Theorem 40 to a set of IF-THEN rules, the result remains the same since in accordance to Theorem 40, replacing μ by its extensional hull does not change the fuzzy set $(\mu \circ_{\sqcap} R^i_*)$ so that neither (4.30) nor (4.31) is affected. The result of Theorem 41 remains also valid for a set of fuzzy IF-THEN rules, i.e. the output fuzzy set is extensional, since, due to Theorem 37, infima and suprema maintain extensionality. The only important thing is that one has to consider fuzzy equivalence relations S (T) on X (Y) such that all the fuzzy set μ_{A_i} (μ_{B_i}) are extensional with respect to S (T). Of course, the most interesting fuzzy equivalence relations are the coarsest ones since they characterize the indistinguishability inherent to the given fuzzy set and yield the greatest extensional hulls. The coarsest fuzzy equivalence relation making a given collection of fuzzy sets extensional is described in Theorem 36.

Example 2 *Fuzzy partition* is a collection of fuzzy sets $\{\mu_i \mid i \in \mathcal{Z}\}$ where, for any real number x, $\mu_i(x) = 1 - \min\{|x - i|, 1\}$ (see Figure 1.) An often applied approximate reasoning scheme for IF-THEN rules is such that L is the Brouwer structure and a fuzzy equality relation is induced by fuzzy partition, that is

$$
\begin{aligned}
S(x,y) &= \bigwedge_{i \in \mathcal{Z}} [\mu_i(x) \leftrightarrow \mu_i(y)] \\
&= \begin{cases} 1 & \text{if } x = y, \\ \min\{i+1-x, x-i, i+1-y, y-i\} & \text{if } i < x, y < i+1, \\ 0 & \text{otherwise.} \end{cases}
\end{aligned}
$$

We visualize a fuzzy partition by the following graph

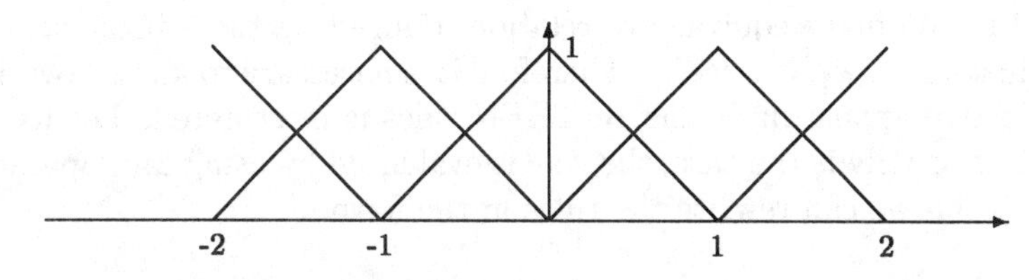

A typical fuzzy partition.

The following figure illustrates the extensional hulls of the crisp values 0.5 and 1.75 with respect to this fuzzy equivalence relation, i.e. the fuzzy sets $\mu_{0.5}(x) = S(x, 0.5)$ and $\mu_{1.75}(x) = S(x, 1.75)$. The greatest indistinguishability is reached at the intermediate points $z + 0.5, z \in \mathcal{Z}$ whereas the extensional hull of the points $z \in \mathcal{Z}$ remains crisp leading to optimal accuracy.

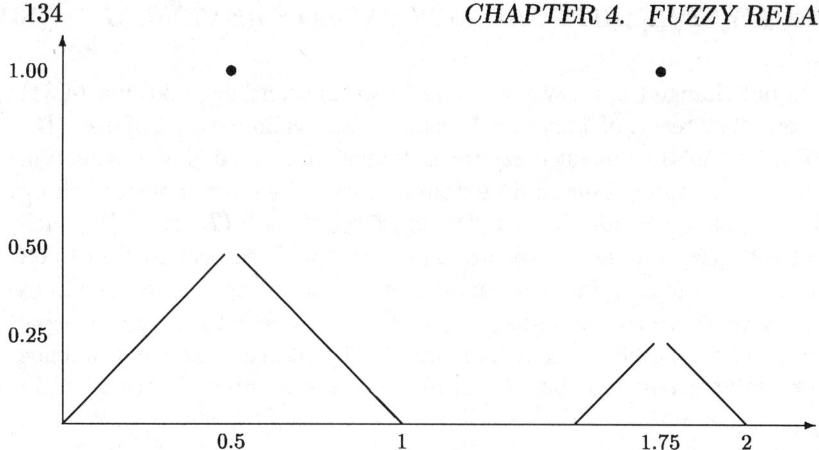

The extensional hulls of the crisp values 0.5 and 1.75.

It should be emphasized that the choice of Brouwer structure, i.e. $\odot =$ min leads to the greatest distinguishability since $\leftrightarrow_{\min} \leq \leftrightarrow_t$ holds for any continuous t-norm, as we show in Exercise 1.4.3.

The IF-THEN rules we have considered till now have had only one variable in the premise and in the conclusion. Nevertheless, these variables may be multi-dimensional vectors, too, thus we can assume multiple inputs and outputs as well.

In such cases the corresponding fuzzy equivalence relations are defined on multi-dimensional product spaces. In some cases it is, however, possible to compute a lower approximation of a fuzzy equivalence relation in a product space by fuzzy equivalence relations in one-dimensional spaces.

Let us assume the IF-THEN rules are of the form

$$\text{if } x_1 \text{ is } A_i^{(1)} \text{ and } \ldots \text{ and } x_n \text{ is } A_i^{(n)}, \text{ then } y \text{ is } B_i, \text{ where } i \in \Gamma,$$

with $X_1 \ldots X_n$ as the underlying domains of the variables $x_1 \ldots x_n$. The linguistic terms $A_i^{(k)}$ are modeled by fuzzy sets $\mu_i^{(k)} \in L^{X_k}$. We are interested in the fuzzy equivalence relation induced by these fuzzy sets on the product space $X_1 \times \ldots \times X_n$. Therefore it is necessary to know how the **and**-connective appearing in the IF-THEN rules is interpreted. Let us assume that, as usually is the case, the rule is evaluated by using min-operation for **and**. Then we can rewrite the rules in the form

$$\text{if } (x_1, \ldots x_n) \text{ is } (A_i^{(1)}, \ldots A_i^{(n)}), \text{ then } y \text{ is } B_i, \text{ where } i \in \Gamma,$$

and the linguistic term $(A_i^{(1)}, \ldots A_i^{(n)})$ is represented by the fuzzy set

$$(\bigwedge_{k=1}^{n} \mu_i^{(k)}) \in L^{X_1 \times \ldots \times X_n}.$$

The corresponding fuzzy equivalence relation on the product space $X_1 \times \ldots \times X_n$ based on formula (4.13) is

$$S((x_1,\ldots,x_n),(x_1',\ldots,x_n')) = \bigwedge_{i\in\Gamma}\{[\bigwedge_{k=1}^{n} \mu_i^{(k)}(x_k)] \leftrightarrow [\bigwedge_{k=1}^{n} \mu_i^{(k)}(x_k')]\}. \quad (4.32)$$

The following theorem shows that a lower approximation for this fuzzy equivalence relation can be derived from the fuzzy equivalence relations in the one-dimensional spaces.

Theorem 42 *Let* $\mu_i^{(k)} \in L^{X_k}$, *where* $i \in \Gamma$ *and* $k = 1,\ldots,n$. *Let* S *be a fuzzy equivalence relation defined by* (4.32). *For all* $x, y \in X_k$ *define*

$$S_k(x,y) = \bigwedge\nolimits_{i\in\Gamma}[\mu_i^{(k)}(x)] \leftrightarrow [\mu_i^{(k)}(x')].$$

Moreover, let

$$\widetilde{S}((x_1,\ldots,x_n),(x_1',\ldots,x_n')) = \bigwedge_{k=1}^{n} S_k(x_k,x_k').$$

Then $\widetilde{S}$ *is a fuzzy equivalence relation on* $X_1 \times \ldots \times X_n$ *and* $\widetilde{S} \leq S$.

Proof. The fuzzy sets $\mu_i^{(k)} \in L^{X_k}$ can be interpreted as fuzzy sets $\widetilde{\mu}_i^{(k)} \in L^{X_1\times\ldots\times X_n}$ by defining

$$\widetilde{\mu}_i^{(k)}(x_1\ldots x_n) = \mu_i^{(k)}(x_k).$$

Thus $\widetilde{S}$ can be rewritten in the form

$$\widetilde{S}((x_1,\ldots,x_n),(x_1',\ldots,x_n')) = \bigwedge_{i\in\Gamma}\bigwedge_{k=1}^{n}\{\widetilde{\mu}_i^{(k)}(x_1\ldots x_n) \leftrightarrow \widetilde{\mu}_i^{(k)}(x_1'\ldots x_n')\}$$

and is therefore, by Theorem 36, a fuzzy equivalence relation on $X_1 \times \ldots \times X_n$. Moreover, $\widetilde{S} \leq S$ holds by (1.60).□

We have seen that fuzzy equivalence relations based on BL-algebras are a useful model to describe the indistinguishability inherent to fuzzy sets. Approximate reasoning schemes that are used in fuzzy control and other fields cannot avoid or overcome this indistinguishability. Thus fuzzy equivalence relations characterize a kind of granularity of the model. This information can be used to determine a limit for the degree of precision in which inputs should be measured. Finally, let us remark on the so-called *defuzzification* problem the following: although defuzzification aims at deleting the indistinguishability or imprecision inherent to a fuzzy set, the results above tell us the indistinguishability inherent to a given collection of fuzzy sets cannot be overcome by any defuzzyfication strategy as defuzzification is applied to the final output fuzzy set of an approximate reasoning scheme and this fuzzy set remains the same even if the input fuzzy set is replaced by its extensional hull.

In the next section, which will be also the last section of this book, we introduce an alternative way to handle fuzzy inference. Our method does not require any defuzzication but is based on a simple notion of *total similarity* which is composed of partial similarities, each generated by a single fuzzy set.

4.4 Maximal Similarity and Fuzzy Reasoning

The aim of this last section is to give an alternative approach to fuzzy reasoning, a method which does not require any defuzzification technique but is based only on an experts knowledge and on the concept of fuzzy similarity. The method we introduce here is rather simple and can be described by the following illustration.

Assume we want to buy a personal computer (PC) which should *not* be *too expensive*, which should have *enough memory* and which *screen* should be *clear enough*. All these three criteria are fuzzy; for sure there is a price which is definitely to high as well as there is a reasonable price, however, there is gray area, a fuzzy set, between these two extremes. The shape of this fuzzy set depends on our choice. The same concerns the memory and the screen of the PC, too. It may happen that none of the PCs available on market fulfill totally all our demands but still we are able choose the best one(s). This is because we have in mind an intuitive concept of a *total similarity* which depends on some *partial similarities*, in this case the partial similarities generated by the above three fuzzy sets, so we choose that PC which has the highest degree of total similarity with the ideal one. Now we show this method can be carried out formally. Our method allows various weight coefficients for various criteria, too.

Proposition 102 *Consider n Łukasiewicz valued fuzzy similarities S_i, $i = 1, \cdots, n$ on a set X. Then*

$$S\langle x, y\rangle = \tfrac{1}{n}\Sigma_{i=1}^{n} S_i\langle x, y\rangle$$

is a Łukasiewicz valued fuzzy similarity on X.

Proof. As all S_i, $i = 1, \cdots, n$ are reflexive and symmetric so is S. The transitivity of S can be seen in the following way. Let $A = S\langle x, y\rangle \odot S\langle y, z\rangle$. Then

$$\begin{aligned}
A &= (\tfrac{1}{n}\Sigma_{i=1}^{n} S_i\langle x, y\rangle) \odot (\tfrac{1}{n}\Sigma_{i=1}^{n} S_i\langle y, z\rangle) \\
&\le \tfrac{1}{n}(\Sigma_{i=1}^{n} S_i\langle x, y\rangle + \Sigma_{i=1}^{n} S_i\langle y, z\rangle - n) \\
&= \tfrac{1}{n}[(S_1\langle x, y\rangle + S_1\langle y, z\rangle - 1) + \cdots + (S_n\langle x, y\rangle + S_n\langle y, z\rangle - 1)] \\
&= \tfrac{1}{n}[(S_1\langle x, y\rangle \odot S_1\langle y, z\rangle) + \cdots + (S_n\langle x, y\rangle \odot S_n\langle y, z\rangle)] \\
&\le \tfrac{1}{n}(S_1\langle x, z\rangle + \cdots + (S_n\langle x, z\rangle) \\
&= S\langle x, z\rangle,
\end{aligned}$$

and the proof is complete.

It will be an exercise for the reader to show that Proposition 102 does not hold for other BL-algebras that MV-algebras.

Remark 27 *In case we want to give different weights to various partial similarities, say* 50%, 30% *and* 20% *in our PC problem above, we just assume there are* 50 *partial similarities of type* 1, 30 *of type* 2 *and* 20 *of type* 3, *thus summing up to* 100 *partial similarities, and put* $n = 100$. *More generally, if we have* $m \leq 100$ *partial similarities* S_i, $i = 1, \cdots m$ *and we want to give them* $p_i\%$ *weights, where* $p_i \in N$, $\Sigma_{i=1}^{m} p_i = 100$, *we multiply each* $S_i\langle x, y\rangle$ *by a factor* p_i *and let* $n = 100$.

4.4.1 Fuzzy Traffic Signals

Consider the following signalized pedestrian crossing[2].

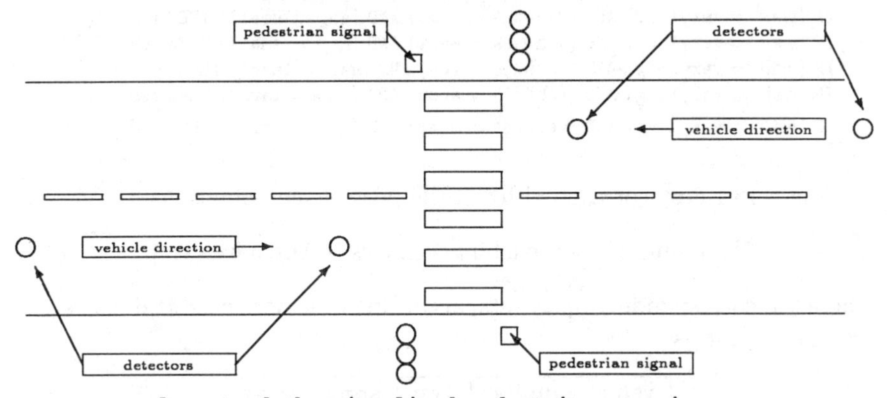

Layout of the signalized pedestrian crossing

The signal operates as follows: when no pedestrian wishing to cross the street is present, the signal phase is green for the vehicular traffic. When a pedestrian arrives, and presses the button, the controller evaluates the state of vehicle arrivals and the waiting time of the pedestrian. If specified criteria are met, the current phase is terminated and a green phase is provided for the pedestrians; otherwise the current phase continues. The minimum green time for roadway is set to 5 seconds and the minimum and the maximum green times for pedestrians are 10 and 30 seconds, respectively. The controller has a multiobjective approach: (1) Minimum pedestrian waiting time[3], (2) Minimum delay to vehicular movement, and (3) Maximum safety to the vehicles and pedestrians[4].

[2]This expert system has been constructed at Helsinki Univ. of Technology, Transportation Engineering Lab.

[3]It has been observed that after about 30 seconds people start to doubt that the signals are broken and they may try to cross the street against red signal thus causing a danger.

[4]When a group of vehicles are approaching, then the whole group should be preserved and passed. This is to avoid rear end collisions, and to great a gap in the downstream flow to provide better opportunities for the pedestrians at downstream crossing to cross.

Experienced traffic signal designers describe possible choices for drivers green signal or red signal by the following fuzzy IF – THEN rules

IF (wait = short) AND (vehicles = some) AND (gap = small) THEN green
IF (wait = short) AND (vehicles = many) AND (gap = small) THEN green
IF (wait = long) AND (vehicles = many) AND (gap = small) THEN green
IF (wait = short) AND (vehicles = few) AND (gap = large)) THEN green
IF (wait = long) AND (vehicles = few) AND (gap = large) THEN green
IF (wait = short) AND (vehicles = some) AND (gap = large) THEN green
IF (wait = long) AND (vehicles = some) AND (gap = large) THEN green
IF (wait = short) AND (vehicles = many) AND (gap = large) THEN green
IF (wait = long) AND (vehicles = many) AND (gap = large) THEN green
IF (wait = short) AND (vehicles = few) AND (gap = small) THEN red
IF (wait = long) AND (vehicles = few) AND (gap = small) THEN red
IF (wait = verylong) AND (vehicles = few) AND (gap = small) THEN red
IF (wait = long) AND (vehicles = some) AND (gap = small) THEN red
IF (wait = verylong) AND (vehicles = some) AND (gap = small) THEN red
IF (wait = verylong) AND (vehicles = many) AND (gap = small) THEN red
IF (wait = verylong) AND (vehicles = few) AND (gap = large) THEN red
IF (wait = verylong) AND (vehicles = some) AND (gap = large) THEN red
IF (wait = verylong) AND (vehicles = many) AND (gap = large) THEN red

These 18 rules[5] correspond to all the possible combinations

Wait time × Approaching Vehicles × Discharge Gap

where the corresponding fuzzy sets, according to experienced traffic designers, are as follows

vehicles [units]	few	some	many
0	1	0	0
1	0.5	0.5	0
2	0	1	0.5
3	0	1	1
4	0	0. 5	1
≥ 5	0	0	1

(discharge) gap [sec]	small	large
1	1	0
1.5	0.8	0.2
2	0.5	0.5
2.5	0.2	0.8
≥ 3	0	1

[5]In general, it is not necessary to have all possible rule combinations; this is one of the big advantages of fuzzy reasoning.

wait (time) [sec]	short	long	verylong
≤ 3	1	0	0
4	0.7	0	0
5	0.5	0.05	0
6	0.4	0.1	0
7	0.3	0.15	0
8	0.2	0.2	0
9	0.15	0.3	0
10	0.1	0.35	0
11	0.05	0.45	0
12	0	0.5	0
13	0	0.6	0.05
14	0	0.75	0.1
15	0	0.85	0.15
16	0	1	0.2
17	0	0.85	0.3
18	0	0.75	0.35
19	0	0.6	0.4
20	0	0.5	0.5
21	0	0.45	0.6
22	0	0.35	0.75
23	0	0.3	0.9
24	0	0.2	1
25	0	0.15	1
26	0	0.1	1
27	0	0.05	1
≥ 28	0	0	1

Besides, an extra rule

In a fifty-fifty situation, extend drivers green signal

is needed. Moreover, traffic designers put weight coefficients 3, 2 and 1 to pedestrian wait time, approaching vehicles and discharge gap, respectively. In practice, the systems measures an actual traffic situation x once per second, compares it with 18 ideal situations $a_1, \cdots, a_{18}$ corresponding to the IF parts of each IF – THEN rule, and then fires that THEN part with which it has the highest degree of total similarity.

Example 1 Let x correspond to a traffic situation such that a pedestrian has been waiting for 15 seconds, there is one vehicle approaching and the discharge gap between this vehicle and the previous one is 4 seconds. The corresponding membership degrees are as follows

$\mu_{\text{short}}(\text{wait} = 15) = 0$, $\mu_{\text{long}}(\text{wait} = 15) = 0.85$, $\mu_{\text{verylong}}(\text{wait} = 15) = 0.15$,
$\mu_{\text{few}}(\text{vehicles} = 1) = 0.5$, $\mu_{\text{some}}(\text{vehicles} = 1) = 0.5$, $\mu_{\text{many}}(\text{vehicles} = 1) = 0$,
$\mu_{\text{small}}(\text{gap} = 4) = 0$, $\mu_{\text{large}}(\text{gap} = 4) = 1$.

When counting total similarities with all the 18 IF-parts, we learn that x has the largest maximal similarity with case
(wait = long) AND (vehicles = few) AND (gap = large)
as well as with case
(wait = long) AND (vehicles = some) AND (gap = large),
namely

$$S(x,a) = \frac{1}{6}[3 \cdot 0.85 + 2 \cdot 0.5 + 1 \cdot 1] = 0.758.$$

In both cases the THEN-part is green.

Example 2 Let x be another traffic situation where a pedestrian has been waiting for 12 seconds, there are three vehicle approaching and the shortest discharge gap between them is 1 second. Now the corresponding membership degrees are
$\mu_{\text{short}}(\text{wait} = 12) = 0$, $\mu_{\text{long}}(\text{wait} = 12) = 0.5$, $\mu_{\text{verylong}}(\text{wait} = 12) = 0$, $\mu_{\text{few}}(\text{vehicles} = 3) = 0$, $\mu_{\text{some}}(\text{vehicles} = 3) = 1$, $\mu_{\text{many}}(\text{vehicles} = 3) = 1$, $\mu_{\text{small}}(\text{gap} = 1) = 1$, $\mu_{\text{large}}(\text{gap} = 1) = 0$.
The largest maximal similarity that x has with is an IF-part
(wait = long) AND (vehicles = some) AND (gap = small)
and an IF-part
(wait = long) AND (vehicles = many) AND (gap = small),
as we calculate

$$S(x,a) = \frac{1}{6}[3 \cdot 0.5 + 2 \cdot 1 + 1 \cdot 1] = 0.75.$$

In the first case the THEN-part is green, in the second case the corresponding THEN-part is red, so we apply the extra rule and determine that drivers signal will be green.

4.4.2 Determining Athlete's Anaerobic Thresholds

The maximal performance capacity is essential in many sports like weight-lifting, while in some other sports like long distance running the submaximal endurance capacity play a more important role. At low exercise levels energy is yielded mostly aerobically, but when approaching maximal exercise level, the aerobic process with increasing lactate production start to play a more perceptible role. To guide successfully athlete's training programs, it is therefore of importance to be able to identify his aerobic and anaerobic thresholds, which are functions of blood lactate, ventilation and oxygen uptake. The test protocol of a continuous incremental exercise, which is performed e.g. by bicycle ergometer, starts with a 3 minutes warm up, then the load is increased every second minute and blood lactate, ventilation and oxygen uptake are measured. The planned duration of the test is about 20-25 minutes, the test is carried out until volitional exhaustion so usually there $x_1, \cdots, x_n$ measurements, where $n = 10 \cdots 12$. Here is a part of a possible test result:

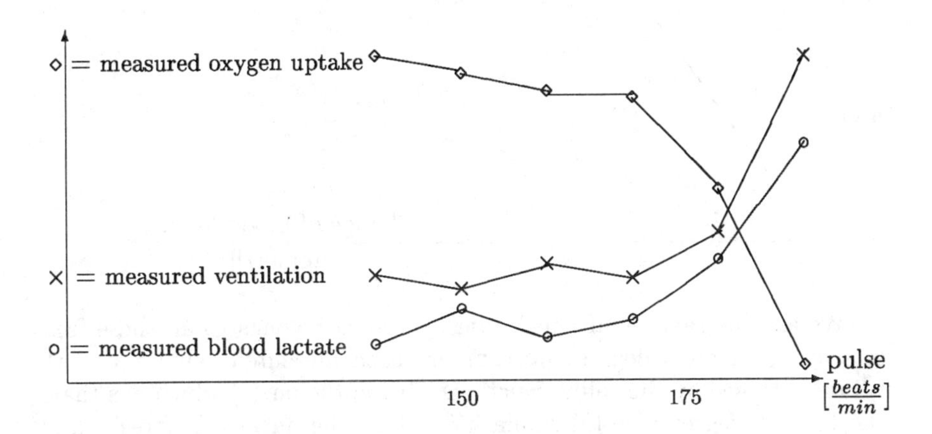

According to skilled sports medicine specialists, the *anaerobic threshold* is such a pulse (beats/minute) that

- the amount of blood lactate is increasing rapidly (more that 0.2 mmol/l)
- ventilation is increasing clearly
- oxygen uptake is decreasing
- pulse is 15-25 (+-5) beats/minute less than maximal pulse.

For example, in the case above, the anaerobic threshold would be about 170 beats/minute. We construct[6] an expert system based on total fuzzy similarity to imitate sports medicine specialists' reasoning. First we connect the measured values x_i with lines as done above (in fact, they are first order spline functions!).

Then, corresponding to each test, we create a fuzzy set called

`Ventilation is increasing clearly`

such that the x_i possessing the absolutely highest positive change of ventilation will have the membership degree 1, the x_j with the absolutely lowest positive change or non-positive change will have the membership degree 0 and all the other x_i:s with positive change will have a linearly scaled degree of membership in this fuzzy set. Due to the spline function, also the values in between measured ones will have a reasonable degree of membership. In a similar manner we define another fuzzy set called

`Oxygen uptake is decreasing.`

The last fuzzy set we need is called

`The amount of blood lactate is increasing rapidly`

and is context independent. It has the following shape

[6] A prototype can be found in [17]

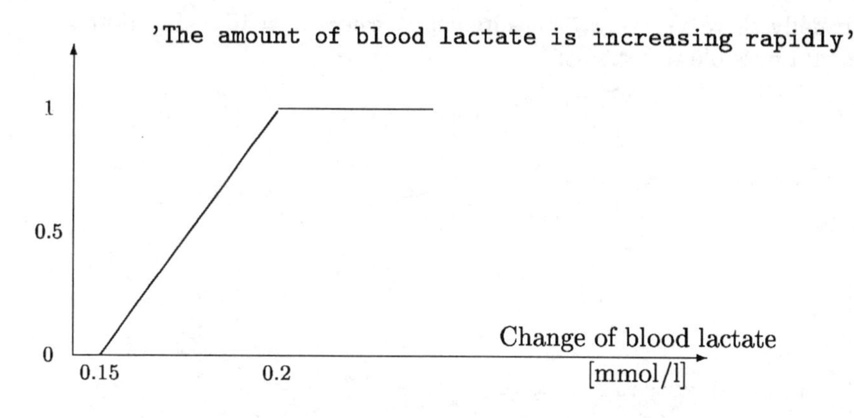

We assume there is an ideal object a which belongs to all these fuzzy sets at the highest degree; we compare each measured value x and the values obtained by the spline functions with a; the one(s) which has (have) the highest degree of total similarity will be the anaerobic threshold(s)[7] if it (they) fulfils the last crisp criteria 'pulse is 10 to 30 beats/minute less than maximal pulse'. It may happen that even the highest degree of total similarity may be absolutely too low, or that there are too many possible anaerobic thresholds. This indicates that something went wrong and the test should be repeated. Many other expert systems do not have this advantage, they simply spit out some result even if it would not make any sense.

Example Having the following test results, define the anaerobic threshold

Pulse	Δ blood lactate	Δ ventilation	Δ oxygen uptake
142	0.14	+21	+0.1
147	0.13	-3	+0.1
156	0.19	-2	-0.1
161	0.20	+1	-0.2
169	0.29	+27	-0.7
174	0.30	+26	-0.7
181	0.35	+26	-0.5
186	0.37	+27	-0.6

For simplicity, we consider only the measured values and calculate the corresponding fuzzy sets and the degrees of total fuzzy similarity. In this test the measured maximal pulse is 186 beats/min so, to fulfill the crisp rule 'pulse is 10 to 30 beats/minute less than maximal pulse', we do not have to consider all the cases. We obtain the following fuzzy sets and results

[7]For *aerobic threshold* there is another rule base

Pulse	Blood Lac.inc.	Vent.inc.	Oxygen Up.inc.	Tot.sim.
156	0.8	0	0	0.27
161	1	0.04	0.17	0.40
169	1	1	1	1
174	1	0.96	1	0.97

The maximal fuzzy similarity is obtained at the value 169 beats/min, while the value 174 beats/min is quite good, too. Since there is always uncertainty involved in the measurements, sports medicine specialists would drive into a conclusion: Anaerobic threshold is in between 169 beats/min and 174 beats/min.

4.4.3 Classification Tasks

Theory of total fuzzy similarity offers us a powerful tool to handle with partial similar objects. For example, consider the following table of information which was collected from The World Almanac and Book of Facts 1998.

State	GDT	area	pop	bir	mor	life	65	lite	tel	car
Finland	18.2	130,1	5.1	11	5	74	14	100	1.8	2.7
Denmark	21.7	16,6	5.3	12	5	74	15	100	1.6	3.1
Belgium	19.5	11.8	10.2	12	6	74	16	99	2.2	2.4
France	18.7	210	58.0	11	6	75	16	99	1.8	2.4
Italy	18.7	116.3	57.5	10	7	75	17	97	2.3	1.9
Spain	14.3	195.4	39.2	10	6	75	16	97	2.6	2.8
Slovakia	7.2	18.9	5.4	13	11	69	11	100	4.8	5.4
Bulgaria	4.9	42.9	8.7	8	15	67	16	98	3.3	5.4
Romania	4.6	92.0	21.4	10	23	66	13	97	7.6	10.7
Colombia	5.3	440.8	37.4	21	25	70	5	91	10.0	32.5
Tanzania	0.8	364.0	29.5	41	105	40	3	68	328	589
Nepal	1.2	56.8	22.6	37	77	54	3	28	276	-

GDP = per capita GDP \$ 1000
pop = population 10^6
mor = infant mortality/1000 live births
65 = age distrib. % 65+
tel = 1 telephone per x persons
area = area sq.mi
bir = births per 1000 pop
life = life expect. at birth
lite = literacy%
car = 1 car per x persons

We may express the information of this table by fuzzy set in various ways. For example, corresponding to the first column, we may construct a fuzzy set `High GDT` by scaling, i.e.

$$\mu_{\texttt{High GDT}}(\text{state x}) = \frac{(\text{GDT of state x}) - (\text{Lowes GDT})}{(\text{Highest GDT}) - (\text{Lowest GDT})}$$

In a similar manner we construct the fuzzy sets

	`Huge Area`
`High Population`	`High Amount of Births`
`High Infant Mortality`	`Long Life Expectancy`
`Many Old Aged`	`High Literacy`

corresponding to the columns 2-8. For the last two columns it is more reasonable to define a fuzzy set `Telephone Per Person` by

$$\mu_{\texttt{Telephone Per Person}}(\text{state x}) = \frac{1}{\text{1 telephone per x persons in state x}}$$

and similarly a fuzzy set `Car Per Person`. This results

State	GDT	area	pop	bir	mor	life	65	lite	tel	car
Finland	0.82	0.27	0	0.09	0	0.97	0.79	1	0.56	0.37
Denmark	1	0.01	0	0.12	0	0.97	0.86	1	0.63	0.32
Belgium	0.89	0	0.1	0.12	0.01	0.97	0.93	0.99	0.45	0.42
France	0.86	0.46	1	0.09	0.01	1	0.93	0.99	0.56	0.42
Italy	0.86	0.24	0.99	0.06	0.02	1	1	0.96	0.43	0.53
Spain	0.65	0.43	0.64	0.06	0.01	1	0.93	0.96	0.38	0.36
Slovakia	0.31	0.02	0.01	0.15	0.06	0.83	0.57	1	0.21	0.19
Bulgaria	0.20	0.07	0.07	0	0.10	0.77	0.93	0.97	0.30	0.19
Romania	0.18	0.19	0.31	0.06	0.18	0.74	0.71	0.96	0.13	0.09
Colombia	0.22	1	0.61	0.39	0.20	0.86	0.14	0.88	0.10	0.03
Tanzania	0	0.82	0.46	1	1	0	0	0.56	0	0
Nepal	0.02	0.11	0.33	0.88	0.72	0.40	0	0	0	0

Example 1 A quick glance at the table shows us that Romania and Bulgaria are the countries most equal to Slovakia with respect to the given information. But which one of them is more equal with Slovakia and what is the degree of their mutual similarity? By using maximal fuzzy similarities generated by the above fuzzy sets we easily calculate that

$$\text{Similar}\langle\text{Slovakia,Bulgaria}\rangle = 0.9050$$
$$\text{Similar}\langle\text{Slovakia,Romania}\rangle = 0.8380$$

Thus, Bulgaria is more equal to Slovakia that Romania is. Moreover,

$$\text{Similar}\langle\text{Romania,Bulgaria}\rangle = 0.8420.$$

Example 2 Typical features of an *undeveloped country* are low GDP per capita, high rate of birth, high infant mortality, short life expectancy at birth and low literacy percentage. In the light of above facts, which are the three most undeveloped countries? It is reasonable to use the following fuzzy sets

```
Low GDP = (High GDP)*
High Amount of Births
High Infant Mortality
Short Life Expectancy = (Long Life Expectancy)*
Low Literacy =(High Literacy)*
```

and to assume that an 'ideal undeveloped country' belongs to each fuzzy set at the degree 1. Comparing now each state with such a country results

State	Degree of being undeveloped
Finland	0.06
Denmark	0.03
Belgium	0.06
France	0.05
Italy	0.05
Spain	0.09
Slovakia	0.21
Bulgaria	0.23
Romania	0.27
Colombia	0.33
Tanzania	0.89
Nepal	0.84

The result is quite expected.

4.5 Exercises

4.1. Solvability of Fuzzy Relation Equations

Exercise 1 Prove Remark 24.

Exercise 2 Prove Remark 25.

Exercise 3 Prove Theorem 28.

Exercise 4 Study the solvability of the fuzzy relation equation $X\mu S = T$, where the composition is the Łukasiewicz composition and the fuzzy relations S and T are defined via

$S\langle v, w\rangle =$

	w_1	w_2	w_3
v_1	0.2	1.0	0.5
v_2	0.1	0.9	0.2
v_3	0.0	1.0	0.4

$T\langle u,w\rangle =$

	w_1	w_2	w_3
u_1	0.3	0.5	0.1
u_2	0.6	0.3	0.4
u_3	0.1	1.0	0.7

Exercise 5 Let S be as in Exercise 4 and

$R\langle u,v\rangle =$

	v_1	v_2	v_3
u_1	0.9	0.8	0.7
u_2	0.6	0.5	0.4
u_3	0.3	0.2	0.1

Define $T = R\mu S$, where μ is the Gaines product composition.

Exercise 6 What is the largest solution of the fuzzy relation equation $X\mu S = T$, where $T = R\mu S$ and R, S, μ as in Exercise 5?

Exercise 7 Mary is a frequent visitor at the students' medical clinic. Her doctor, however, has realized that Mary's symptoms; headache, nausea, constipation, diarrhea, fever and erythema vary on the scale $[0,1]$. The doctor has diagnosed Mary's 'illnesses'; they are hangover, pregnancy, flu, overexertion or some of their combination. As medical treatment the doctor orders rest, pregnancy test, painkiller or outdoor life, each of them again on the scale $[0,1]$. These facts can be expressed by means of fuzzy relations as follows:

$R \underset{\sim}{\subset}$ SYMPTHOMS $\times$ ILLNESSES $=$

	hangover	pregnancy	flu	overexertion
headache	0.9	0.0	0.7	0.3
nausea	0.2	1.0	0.3	0.0
constipation	0.0	0.0	0.0	1.0
diarrhea	0.4	0.0	0.3	0.0
fever	0.0	0.0	1.0	0.0
erythema	0.2	0.0	0.6	1.0

$S \underset{\sim}{\subset}$ ILLNESSES $\times$ MEDICAL TREATMENT $=$

	rest	preg.test	painkiller	outdoor l.
hangover	1.0	0.0	1.0	0.7
pregnancy	0.5	1.0	0.0	0.5
flu	1.0	0.0	0.8	0.0
overexertion	1.0	0.0	0.1	1.0

Form a fuzzy relation $T = R\mu S \subsetneqq$ SYMPTOMS $\times$ MEDICAL TREATMENT where μ is Łukasiewicz product. What does this fuzzy relation tell us?

Exercise 8 Let T and S be as in Exercise 7. Find the maximal solution X of the fuzzy relation equation $X\mu S = T$. What information does this solution contain?

4.2. On Fuzzy Similarity Relations

Exercise 1 Prove the one-to-one correspondence between pseudo-metrics bounded by 1 and fuzzy equivalence relations with respect to the Łukasiewicz product.

Exercise 2 Replace $d(x,z) \leq \max\{d(x,y), d(y,z)\}$ for $d(x,z) \leq d(x,y) + d(y,z)$ in the definition of pseudo-metric. Then we obtain the definition on *ultra-metric*. Prove, similarly to Exercise 1, the one-to-one correspondence between ultra-metrics bounded by 1 and fuzzy equivalence relations with respect to the Brouwer product.

4.3. Fuzzy Similarity and Approximate Reasoning

Exercise 1 Prove Remark 26.

Exercise 2 Show that Theorem 40 does not, in general hold for (a) the combination $\square = \rightarrow$ and $\sqcap = \wedge$, (b) the combination $\square = \odot$ and $\sqcap = \wedge$, (c) the combination $\square = \sqcap = \wedge$.T

4.4. Maximal Similarity and Fuzzy Reasoning

Exercise 1 Show that Proposition 102 does not hold if one uses (a) Product structure (b) Gödel structure on the real unit interval (hint: find a simple counter example).

Exercise 2 Let x be a traffic situation where pedestrians have been waiting for 25 seconds, there are two vehicle approaching and the discharge gap between them is 4 seconds. What will be the signal for drivers?

Exercise 3 Define the anaerobic threshold on a test resulted the following

Pulse	Δ blood lactate	Δ ventilation	Δ oxygen uptake
135	0.11	+11	+0.1
143	0.15	-5	+0.1
151	0.15	-5	-0
159	0.18	0	-0.1
166	0.19	+28	-0.5
172	0.21	+26	-0.8
180	0.35	+27	-0.6
187	0.40	+30	-0.9
196	0.47	+37	-1.6

Exercise 4 According to the table in Subsection 4.4.3, are Finland and Belgium more similar to each other that Italy and France are to each other?

Chapter 5

Solutions to Exercises

5.1 Exercises from Section 1.1

Exercise 1 No! R is reflexive as everyone understands his/hers own language but R is not transitive; a person x may understand a person's y language and person y may, in turn, understand a person's z language but that does not mean that x would necessarily understand the language of z.

Exercise 2 (a): By (1.1), $x \sim x$, therefore $x \in |x|$.
(b): If $x \in |y|$, then $x \sim y$. Assume $x \sim y$. If $z \in |x|$, then $z \sim x$. By (1.2), $z \sim y$. Thus, $z \in |y|$ and therefore $|x| \subseteq |y|$. Similarly, $|y| \subseteq |x|$, hence $|x| = |y|$. Finally, assume $|x| = |y|$. Then $x \in |x| = |y|$.
(c): Assume $x \not\sim y$ but $z \in |x| \cap |y|$ for some element z. Then $z \sim x$, $z \sim y$. By (1.4), $x \sim z$ and by (1.2) $x \sim y$, a contradiction. Therefore $|x| \cap |y| = \emptyset$.

Exercise 3 (a): The relation R is clearly reflexive and transitive. It is also anti-symmetric (there are no two states with exactly the same amount of inhabitants!) Thus, R is an order and defines a chain.
(b): Again, R is an order relation in the set A of all independent states. The order R is not, however, a chain as e.g. Russia and India are incomparable with respect to this order.

Exercise 4 The set inclusion relation $\subseteq$, defined by $X \subseteq Y$ iff $a \in X$ implies $a \in Y$, has clearly the following properties (i) $X \subseteq X$ (ii) if $X \subseteq Y$, $Y \subseteq Z$, then $X \subseteq Z$ (iii) if $X \subseteq Y$, $Y \subseteq X$, then $X = Y$ for any sets X, Y, Z, in particular, for any sets $X, Y, Z \in E$.

Exercise 5 Since A is a chain either $x \leq y$ or $y \leq x$ holds for any $x, y \in A$. Thus, $x \wedge y$ and $x \vee y$ are always defined. Let $x \leq y$. Then $x \wedge y \leq x, y$. If $z \leq x, y$ for some $z \in A$, then $z \leq x \wedge y$. Thus $x \wedge y = \text{g.l.b.}\{x, y\}$.

Similarly, $x \vee y = \text{l.u.b.}\{x, y\}$. The case $y \leq x$ is symmetric. Therefore A is a lattice.

Exercise 6 (1.10): By (1.1), $x \leq x \wedge x \leq x$. Thus, $x = x \wedge x$. Similarly, $x = x \vee x$.
(1.11): Since $x \wedge y \leq y, x$ we have $x \wedge y \leq y \wedge x$. Similarly $y \wedge x \leq x \wedge y$. Therefore $x \wedge y = y \wedge x$. The proof for $x \vee y = y \vee x$ is similar.
(1.12): Because $x \wedge (y \wedge z) \leq x, y \wedge z$ and $y \wedge z \leq y, z$, we reason by (1.2) that $x \wedge (y \wedge z) \leq x, y, z$. On the other hand, if $w \leq x, y, z$, then $w \leq x, y \wedge z$, thus $w \leq x \wedge (y \wedge z)$ so that $x \wedge (y \wedge z) = \text{g.l.b.}\{x, y, x\}$. Similarly we show that also $(x \wedge y) \wedge z = \text{g.l.b.}\{x, y, x\}$. Therefore $x \wedge (y \wedge z) = (x \wedge y) \wedge z$. The proof for $x \vee (y \vee z) = (x \vee y) \vee z$ is symmetric.
(1.13): Since $x \leq x, x \vee y$, we have $x \leq x \wedge (x \vee y)$. On the other hand $x \wedge (x \vee y) \leq x$, thus $x \wedge (x \vee y) = x$. Equally, $x = x \vee (x \wedge y)$.
(1.14): If $x \leq y$ then, as $x \leq x$, $x \leq x \wedge y \leq x$. Thus, $x = x \wedge y$. Similarly we see that $y = x \vee y$. If $y = x \vee y$, then $x \leq x \vee y = y$ and if $x = x \wedge y$, then $x = x \wedge y \leq y$. Thus, $x \leq y$ iff $x = x \wedge y$ iff $y = x \vee y$.

Exercise 7 By (1.10), (1.14), (1.12), (1.11),we have $x \vee z = (x \vee x) \vee z$ $= (x \vee x) \vee (y \vee z) = (x \vee y) \vee (x \vee z)$. Thus, by (1.14), $x \vee y \leq x \vee z$.

Exercise 8 Assume (1.16) holds. Then

$$\begin{array}{rcll}
x \wedge (y \vee z) & = & x \wedge (z \vee y) & \text{(by (1.11))} \\
& = & [x \wedge (x \vee z)] \wedge (z \vee y) & \text{(by (1.13))} \\
& = & [x \wedge (z \vee x)] \wedge (z \vee y) & \text{(by (1.11))} \\
& = & x \wedge [(z \vee x) \wedge (z \vee y)] & \text{(by (1.12))} \\
& = & x \wedge [z \vee (x \wedge y)] & \text{(by (1.16))} \\
& = & x \wedge [(x \wedge y) \vee z] & \text{(by (1.11))} \\
& = & [x \vee (x \wedge y)] \wedge [(x \wedge y) \vee z] & \text{(by (1.13))} \\
& = & [(x \wedge y) \vee x)] \wedge [(x \wedge y) \vee z] & \text{(by (1.11))} \\
& = & (x \wedge y) \vee (x \wedge z) & \text{(by (1.16))}
\end{array}$$

Exercise 9 Assume $a \in X \cap (Y \cup Z)$. Then $a \in X$, therefore $a \in X \cup Y$, $a \in X \cup Z$, thus $a \in (X \cup Y) \cap (X \cup Z)$. We conclude that $X \cap (Y \cup Z)$ $\subseteq (X \cup Y) \cap (X \cup Z)$. Conversely, assume $a \in (X \cup Y) \cap (X \cup Z)$. Then $a \in (X \cup Y)$ and $a \in (X \cup Z)$, thus, $a \in X$ or $a \in Y$ and $a \in X$ or $a \in Z$, thus, $a \in X$ and $a \in Y$ or $a \in Z$. Hence, $a \in X$ and $a \in (Y \cup Z)$ and finally, $a \in X \cap (Y \cup Z)$. Therefore $(X \cup Y) \cap (X \cup Z) \subseteq X \cap (Y \cup Z)$. Thus (1.15) holds.

Exercise 10 By (1.14), $(x \wedge y) \vee (x \wedge z) = x \wedge y$ or $x \wedge z$. Since $x \wedge y$ $\leq x \wedge (y \vee z)$, $x \wedge z \leq x \wedge (y \vee z)$, we conclude that $(x \wedge y) \vee (x \wedge z) \leq$ $x \wedge (y \vee z)$. Conversely, $x \wedge (y \vee z) = y \vee z$ or x. In the first case $y, z \leq x$, therefore

$y = x \wedge y$, $z = x \wedge z$ and $(x \wedge y) \vee (x \wedge z) = y \vee z$. In the second case, as $x \leq y \vee z$, we have $x \leq y$ or $x \leq z$. Thus, $x = x \wedge y$ or $x = x \wedge z$ and therefore $x \leq (x \wedge y) \vee (x \wedge z)$. In both cases $x \wedge (y \vee z) \leq (x \wedge y) \vee (x \wedge z)$. Therefore (1.15) holds.

Exercise 11 No! Take $x = \frac{1}{2}$, $y = \frac{1}{3}$. Then we have $(x \wedge x^*) \vee y = \max\{\frac{1}{3}, \min\{\frac{1}{2}, 1 - \frac{1}{2}\}\} = \frac{1}{2} \neq y$.

Exercise 12 In a Boolean algebra L holds, for any $x, y \in L$, $(x \wedge x^*) \vee y = y = (x \vee x^*) \wedge y$. Thus, for any $y \in L$, $(x \wedge x^*) \leq y \leq (x \vee x^*)$. Therefore $(x \wedge x^*)$ and $(x \vee x^*)$ are the least elements and the greatest elements in L, respectively.

Exercise 13 Assume y, z are two l-complements of $x \in L$, L a Boolean algebra. Then

$$\begin{aligned} y &= (x \wedge z) \vee y && \text{(by definition)} \\ &= (x \vee y) \wedge (z \vee y) && \text{(by (1.16))} \\ &= \mathbf{1} \wedge (z \vee y) && \text{(by Exercise 12)} \\ &= z \vee y && \text{(by (1.14))} \end{aligned}$$

Thus, $y \leq z$, and by symmetry, $z \leq y$. Therefore $z = y = x^*$.

Exercise 14 In a Boolean algebra L, for all $x, y \in L$,

$$\begin{aligned} (x \wedge y) \wedge (x^* \vee y^*) &= [(x \wedge y) \wedge x^*] \vee [(x \wedge y) \wedge y^*] && (1.15) \\ &= [(x \wedge x^*) \wedge y] \vee [(y \wedge y^*) \wedge x] && (1.11), (1.12) \\ &= (\mathbf{0} \wedge y) \vee (\mathbf{0} \wedge x) && \text{(Exercise 12)} \\ &= \mathbf{0} \vee \mathbf{0} && (1.14) \\ &= \mathbf{0} && (1.10) \end{aligned}$$

and

$$\begin{aligned} (x^* \vee y^*) \vee (x \wedge y) &= [(x^* \vee y^*) \vee x] \wedge [(x^* \vee y^*) \vee y] && (1.16) \\ &= [(x^* \vee x) \vee y^*] \wedge [x^* \vee (y^* \vee y)] && (1.11), (1.12) \\ &= (\mathbf{1} \vee y^*) \wedge (x^* \vee \mathbf{1}) && \text{(Exercise 12)} \\ &= \mathbf{1} \wedge \mathbf{1} && (1.14) \\ &= \mathbf{1} && (1.10) \end{aligned}$$

Thus, for any element $z \in L$,
$[(x \wedge y) \wedge (x^* \vee y^*)] \vee z = \mathbf{0} \vee z = z$, $[(x^* \vee y^*) \vee (x \wedge y)] \wedge z = \mathbf{1} \wedge z = z$. By definition, $x^* \vee y^*$ is the l-complement of $x \wedge y$, i.e. $(x \wedge y)^* = x^* \vee y^*$. Since the l-complement of an element $x^* \in L$ is unique we reason that $(x^*)^* = x$. This implies $(x^* \wedge y^*)^* = (x^*)^* \vee (y^*)^* = x \vee y$ and therefore $(x \vee y)^* = x^* \wedge y^*$. Thus, any Boolean algebra is a soft algebra.

Exercise 15 For all real numbers $x, y \in [0, 1]$ holds
$1 - \min\{x, y\} = \max\{1 - x, 1 - y\}$, $1 - \max\{x, y\} = \min\{1 - x, 1 - y\}$. Therefore $\langle I, \leq, \min, \max, ^* \rangle$ is a soft algebra.

5.2 Exercises from Section 1.2

Exercise 1 (1.19) trivially implies (1.20). Assume $a \in F$, $a \leq b$. Then $a = a \wedge b \in F$ thus, by (1.19), $b \in F$. Therefore (1.21) holds. Conversely, let (1.20) and (1.21) hold, $a \wedge b \in F$. Since $a \wedge b \leq a, b$ we have $a, b \in F$. Thus, (1.19) is valid. Let (1.21) hold. If $a \in F$, $b \in L$ then $a \leq a \vee b$, thus $a \vee b \in F$. Conversely, let (1.22) hold. If $a \in F$, $a \leq b$, then $a \vee b = b \in F$.

Exercise 2 (a): Let $a, b \in A$. Then we have $a, b \in G$ iff $h(a) = h(b) = \mathbf{1}_\mathrm{B}$ iff $h(a) \wedge h(b) = \mathbf{1}_\mathrm{B}$ iff $h(a \wedge b) = \mathbf{1}_\mathrm{B}$ iff $a \wedge b \in G$.
(b): Let $a, b \in A$. Then $a, b \in H$ iff $h(a), h(b) \in F$ iff $h(a) \wedge h(b) \in F$ iff $h(a \wedge b) \in F$ iff $a \wedge b \in H$. (Of course, (a) is a special case of (b) as $\{\mathbf{1}_\mathbf{B}\}$ is a filter of B !).

5.3 Exercises from Section 1.3

Exercise 1 Let $x \leq y$. Since μ is isotone we have $xg(y, z) \leq yg(y, z) \leq z$ (as $g(y, z) \leq g(y, z)$). Thus $g(y, z) \leq g(x, z)$. Therefore g is antitone in the first variable. It is also isotone in the second variable as $zg(z, x) \leq x \leq y$, therefore $g(z, x) \leq g(z, y)$.

Exercise 2 Assume (A). Then

$$[(y \to z) \odot (x \to y)] \odot x \leq (y \to z) \odot [(x \to y) \odot x] \leq (y \to z) \odot y \leq z.$$

Hence $(y \to z) \odot (x \to y) \leq x \to z$, whence $y \to z \leq (x \to y) \to (x \to z)$. Thus (A') holds. Assume (A'). Then $b \to c \leq (a \to b) \to (a \to c)$ which implies $[(b \to c) \odot (a \to b)] \odot a \leq c$. Take $a = z$, $b = y \odot z$, $c = x \odot (y \odot z)$. Then

$$\{[(y \odot z) \to x \odot (y \odot z)] \odot [z \to (y \odot z)]\} \odot z \leq x \odot (y \odot z).$$

Now $x \leq (y \odot z) \to [x \odot (y \odot z)]$ and $y \leq z \to (y \odot z)$. Thus,

$$(x \odot y) \odot z \leq \{[(y \odot z) \to x \odot (y \odot z)] \odot [z \to (y \odot z)]\} \odot z \leq x \odot (y \odot z).$$

Therefore (A) holds.

Assume (B). Since $\to$ is antitone in the first variable we have, for each $j \in \Gamma$, $\bigvee_{i \in \Gamma} y_i \to x \leq y_j \to x$, thus $\bigvee_{i \in \Gamma} y_i \to x \leq \bigwedge_{i \in \Gamma}(y_i \to x)$. Conversely, for any $j \in \Gamma$, $\bigwedge_{i \in \Gamma}(y_i \to x) \leq y_j \to x$, so $\bigwedge_{i \in \Gamma}(y_i \to x) \odot y_j \leq x$, thus $\bigvee_{j \in \Gamma}[\bigwedge_{i \in \Gamma}(y_i \to x) \odot y_j] \leq x$. By (B), $\bigwedge_{i \in \Gamma}(y_i \to x) \odot \bigvee_{j \in \Gamma} y_j \leq x$. Thus $\bigwedge_{i \in \Gamma}(y_i \to x) \leq \bigvee_{j \in \Gamma} y_j \to x$. We conclude that (B') holds. Let now (B') hold. Since for each $j \in \Gamma$, $x \odot y_j \leq \bigvee_{i \in \Gamma}(x \odot y_i)$ we conclude, for each $j \in \Gamma$, that $x \leq y_j \to \bigvee_{i \in \Gamma}(x \odot y_i)$. Therefore

$$x \leq \bigwedge_{j \in \Gamma}[y_j \to \bigvee_{i \in \Gamma}(x \odot y_i)].$$

Thus, by (B'), $x \leq \bigvee_{j \in \Gamma} y_j \to \bigvee_{i \in \Gamma}(x \odot y_i)$. Therefore

$$x \odot \bigvee_{i\in\Gamma} y_i \leq \bigvee_{i\in\Gamma}(x \odot y_i).$$

On the other hand, for each $i \in \Gamma$, $x \odot y_i \leq x \odot \bigvee_{i\in\Gamma} y_i$, hence $\bigvee_{i\in\Gamma}(x \odot y_i) \leq x \odot \bigvee_{i\in\Gamma} y_i$. We conclude that (B) holds. Let (C) hold. By the Galois correspondence and since $\to$ is isotone in the second variable, we have $x \leq \mathbf{1} \to x \leq \mathbf{1} \to (\mathbf{1} \to x)$ which implies $[(\mathbf{1} \to x) \odot \mathbf{1}] \odot \mathbf{1} \leq x$. Thus, by (C), $\mathbf{1} \to x \leq x$. We conclude that (C') holds. Assume (C'). Then $x \odot \mathbf{1} \leq x$. On the other hand, $(\mathbf{1} \to x) \odot \mathbf{1} \leq x \odot \mathbf{1}$, therefore $\mathbf{1} \to x \leq \mathbf{1} \to (x \odot \mathbf{1})$, thus $x \leq x \odot \mathbf{1}$ and so (C) holds.

Let (D) hold. Then $x \leq y$ iff $\mathbf{1} \odot x \leq y$ iff $\mathbf{1} \leq x \to y$ iff $\mathbf{1} = x \to y$, i.e. (D'). Conversely, assume (D'). Since $x \leq x$, $\mathbf{1} = x \to x$ so that $\mathbf{1} \odot x \leq x$. On the other hand, $\mathbf{1} \odot x \leq \mathbf{1} \odot x$, thus $\mathbf{1} \leq x \to (\mathbf{1} \odot x) \leq \mathbf{1}$, hence $\mathbf{1} = x \to (\mathbf{1} \odot x)$ which implies $x \leq x \odot \mathbf{1}$. Therefore (D) is valid.

Let (E) hold. Then $x \leq y \to z$ iff $x \odot y \leq z$ iff $y \odot x \leq z$ iff $y \leq x \to z$. Thus (E'). Assume now (E'). Since $y \odot x \leq y \odot x$ we have $y \leq x \to (y \odot x)$, by (E'), $x \leq y \to (y \odot x)$. Thus $x \odot y \leq y \odot x$. By a similar argument we see that also $y \odot x \leq x \odot y$. Therefore (E).

Assume (F). Now $(x \to y) \odot x \leq y$. By (F) and since $\odot$ is isotone, it holds that $(x \to y) \odot (x \odot z) \leq [(x \to y) \odot x] \odot z \leq y \odot z$. Thus, $x \to y \leq (x \odot z) \to (y \odot z)$. Hence (F') is valid. Conversely, let (F') hold. Since $x \leq y \to (x \odot y)$, we have $x \leq (y \odot z) \to [(x \odot y) \odot z]$ and therefore $x \odot (y \odot z) \leq [(x \odot y) \odot z]$, i.e. (F).

Let (G) hold. Then $\{[(x \odot y) \to z] \odot x\} \odot y = [(x \odot y) \to z] \odot (x \odot y) \leq z$. Therefore $[(x \odot y) \to z] \odot x \leq y \to z$ and so $(x \odot y) \to z \leq x \to (y \to z)$. Conversely, by (G), Galois correspondence and as $\odot$ is isotone,

$$[x \to (y \to z)] \odot (x \odot y) = \{[x \to (y \to z)] \odot x\} \odot y \leq (y \to z) \odot y \leq z.$$

Therefore $x \to (y \to z) \leq (x \odot y) \to z$. We conclude that (G') is valid. Finally, assume (G'). Since $x \odot (y \odot z) \leq x \odot (y \odot z)$, we have

$$x \leq (y \odot z) \to [x \odot (y \odot z)]$$

implying $x \leq y \to \{z \to [x \odot (y \odot z)]\}$, thus $x \odot y \leq z \to [x \odot (y \odot z)]$ and therefore $(x \odot y) \odot z \leq x \odot (y \odot z)$. Conversely, $x \odot (y \odot z) \leq (x \odot y) \odot z$ iff $x \leq (y \odot z) \to [(x \odot y) \odot z]$ iff $x \leq y \to \{z \to [(x \odot y) \odot z]\}$ iff $x \odot y \leq z \to [(x \odot y) \odot z]$ iff $(x \odot y) \odot z \leq (x \odot y) \odot z$, which is the case. Consequently, (G) holds.

Exercise 3 (a): $x \odot y \leq z$ iff $x \leq \bigvee\{w \mid w \odot y \leq z\} = y \to z$, thus (1.27) holds.

(b): Let $w \leq x$. Since $\odot$ is isotone, $z \odot w \leq y$ for any such $z \in L$ that $z \odot x \leq y$. Therefore $\bigvee\{z \odot w \mid z \odot x \leq y\} \leq y$. By (1.28), $\bigvee\{z \mid z \odot x \leq y\} \odot w \leq y$, where, by (1.29), $x \to y = \bigvee\{z \mid z \odot x \leq y\}$. Thus, $(x \to y) \odot w \leq y$. By the Galois correspondence (1.27), $x \to y \leq w \to y$. Therefore $\to$ is antitone in the first variable.

(c): If $x \leq y$, then $z \to x = \bigvee\{w \mid w \odot z \leq x\} \leq \bigvee\{w \mid w \odot z \leq y\} = z \to y$.

Exercise 4 (a): Let $x \leq y$. Since $\to$ is isotone in the second variable we have $x \odot w = \bigwedge\{z \mid x \leq w \to z\} \leq \bigwedge\{z \mid y \leq w \to z\} = y \odot w$. Therefore $\odot$ is isotone in the first variable. If z is such that $w \leq y \to z$, then $w \leq x \to z$ as $\to$ is antitone in the first variable. Therefore $w \odot x = \bigwedge\{z \mid w \leq x \to z\}$ $\leq \bigwedge\{z \mid w \leq y \to z\} = w \odot y$. Therefore $\odot$ is isotone in the second variable, too.
(b): Since $\to$ is isotone in the second variable, $x \odot y = \bigwedge\{w \mid x \leq y \to w\}$ $\leq z$ iff $x \leq y \to z$.

Exercise 5 Assume $x, y, z \in I$. $\odot$ is commutative as

$$\begin{aligned} x \odot y &= 1 - \min\{1, \sqrt[p]{(1-x)^p + (1-y)^p}\} \\ &= 1 - \min\{1, \sqrt[p]{(1-y)^p + (1-x)^p}\} \\ &= y \odot x. \end{aligned}$$

$\odot$ is isotone, indeed, if $x \leq y$, then $(1-y)^p \leq (1-x)^p$ and

$$\begin{aligned} x \odot z &= 1 - \min\{1, \sqrt[p]{(1-x)^p + (1-z)^p}\} \\ &\leq 1 - \min\{1, \sqrt[p]{(1-y)^p + (1-z)^p}\} \\ &= y \odot z. \end{aligned}$$

$\odot$ is associative; first let $(1-x)^p + (1-y)^p + (1-z)^p < 1$. Then
$x \odot (y \odot z)$

$$\begin{aligned} &= 1 - \min\{1, \sqrt[p]{(1-x)^p + [1-(1-\min\{1, \sqrt[p]{(1-y)^p + (1-z)^p}\})]^p}\} \\ &= 1 - \min\{1, \sqrt[p]{(1-x)^p + (1-y)^p + (1-z)^p}\} \\ &= 1 - \min\{1, \sqrt[p]{[1-(1-\min\{1, \sqrt[p]{(1-x)^p + (1-y)^p}\}]^p + (1-z)^p)}\} \\ &= (x \odot y) \odot z. \end{aligned}$$

On the other hand, if $1 \leq (1-x)^p + (1-y)^p + (1-z)^p$, then

$$x \odot (y \odot z) = 1 - \min\{1, \sqrt[p]{(1-x)^p + \min\{1, (1-y)^p + (1-z)^p\}}\} = 0$$

and

$$(x \odot y) \odot z = 1 - \min\{1, \sqrt[p]{\min\{1, (1-x)^p + (1-y)^p\} + (1-z)^p}\} = 0.$$

Thus, in both cases $x \odot (y \odot z) = (x \odot y) \odot z$.

We also have

$$\begin{aligned} x \odot 1 &= 1 - \min\{1, \sqrt[p]{(1-x)^p + (1-1)^p}\} \\ &= 1 - \min\{1, (1-x)\} \\ &= x. \end{aligned}$$

For $p = 1$, this structure coincides with Łukasiewicz structure. Indeed, in both structures $x \to y = \min\{1, 1-x+y\}$ and

$$\begin{aligned} x \odot y &= 1 - \min\{1, (1-x) + (1-y)\} \\ &= 1 - \min\{1, 2 - x - y\} \\ &= \max\{0, x + y - 1\}. \end{aligned}$$

Exercise 6 The min-operation is trivially associative, commutative and isotone in the unit interval and $\min\{1, x\} = x$. Let $\min\{x, y\} \leq z$. If $x \leq y$, then $x = \min\{x, y\} \leq z \leq y \to z$ and, if $y \leq x$, then $y = \min\{x, y\}$ and $x \leq 1 = y \to z$. Assume now $x \leq y \to z$. If $y \to z = 1$ then $y \leq z$ and $\min\{x, y\} \leq z$. If $y \to z = z$, then $x \leq z$, thus $\min\{x, y\} \leq z$. We conclude that the Galois correspondence holds. Therefore Brouwerian structure is a residuated lattice.

Exercise 7 The multiplication of real numbers is obviously associative, commutative and isotone and $1 \cdot x = x$. Moreover, if $y \leq z$, then trivially $x \cdot y \leq z$ iff $x \leq y \to z = 1$ and if $y > z$, then $x \cdot y \leq z$ iff $x \leq \frac{z}{y} = y \to z$ Thus, Gaines structure is a residuated lattice.

Exercise 8 Let L be a Boolean algebra. The operation $\wedge$ is associative, commutative, isotone and, for any $x \in L$, $x \wedge \mathbf{1} = x$. Assume $x \wedge y \leq z$. Then, as Boolean algebras are distributive, $x \leq x \vee y^* = (x \vee y^*) \wedge \mathbf{1} = (x \vee y^*) \wedge (y \vee y^*) = (x \wedge y) \vee y^* \leq z \vee y^* = y^* \vee z$. Conversely, let $x \leq y^* \vee z$. Then $x \wedge y \leq y \wedge (y^* \vee z) = (y \wedge y^*) \vee (y \wedge z) = \mathbf{0} \vee (y \wedge z) = y \wedge z \leq z$. Thus, $x \wedge y \leq z$ iff $x \leq y^* \vee z$. Therefore, by defining $x \to y = x^* \vee y$, L can be considered a residuated lattice.

Exercise 9 (1.34): $\mathbf{1} \to x = \bigvee\{z \mid z \odot \mathbf{1} \leq x\} = \bigvee\{z \mid z \leq x\} = x$.
(1.35): Since $\mathbf{1} \odot x \leq x$, we have $\mathbf{1} \leq x \to x \leq \mathbf{1}$. Therefore $x \to x = \mathbf{1}$.
(1.36): As $x, y \leq \mathbf{1}$, we reason that $x \odot y \leq x \odot \mathbf{1} = x$, $x \odot y \leq \mathbf{1} \odot y = y$.
(1.37): Follows immediately from (1.36).
(1.38): As $x \odot y \leq y$, $y \leq x \to y$.
(1.39): Follows immediately from (1.38). (1.40): $\mathbf{1} \odot x = x \leq y$ iff $\mathbf{1} \leq x \to y \leq \mathbf{1}$.
(1.41): If $x = y$, then, by (1.35), $x \to y = y \to x = \mathbf{1}$. Conversely, if $x \to y = y \to x = \mathbf{1}$, then, by (1.40), $x \leq y$, $y \leq x$, therefore $x = y$.
(1.42): Because $\mathbf{1} \odot x \leq \mathbf{1}$, we have $\mathbf{1} \leq x \to \mathbf{1} \leq \mathbf{1}$.
(1.43): Since $\mathbf{0} = \mathbf{1} \odot \mathbf{0} \leq x$, we have $\mathbf{1} \leq \mathbf{0} \to x \leq \mathbf{1}$.
(1.44): $\mathbf{1} \leq x \to (y \to x)$ iff $x \leq y \to x$, which holds true by (1.38).
(1.45): $\mathbf{1} \leq (x \to y) \to [(y \to z) \to (x \to z)]$ iff $x \to y \leq (y \to z) \to (x \to z)$ iff $(x \to y) \odot (y \to z) \leq x \to z$ iff $[(x \to y) \odot (y \to z)] \odot x \leq z$, which is the case as $(x \to y) \odot (y \to z) \odot x \leq (y \to z) \odot y \leq z$.
(1.46): $\mathbf{1} \leq (x \to y) \to [(z \to x) \to (z \to y)]$ iff $x \to y \leq (z \to x) \to (z \to y)$ iff $(x \to y) \odot (z \to x) \leq z \to y$ iff $[(x \to y) \odot (z \to x)] \odot z \leq y$, which holds true as $(x \to y) \odot (z \to x) \odot z \leq (x \to y) \odot x \leq y$.
(1.48): $x \to (y \to z) = (x \odot y) \to z = (y \odot x) \to z = y \to (x \to z)$.

Exercise 10 In Gödel structure $(\frac{1}{2})^{**} = 1$; in Product structure $(\frac{2}{3}) \to (\frac{1}{3}) = \frac{1}{2}$, while $(\frac{1}{3})^* \to (\frac{2}{3})^* = 1$.

Exercise 11 (1.56): $x \leftrightarrow \mathbf{1} = (x \to \mathbf{1}) \wedge (\mathbf{1} \to x) = \mathbf{1} \wedge x = x$.
(1.57): $x \leftrightarrow y = (x \to y) \wedge (y \to x) = \mathbf{1}$ iff $(x \to y) = (y \to x) = \mathbf{1}$ iff $x = y$.
(1.58): $x \leftrightarrow y = (x \to y) \wedge (y \to x) = (y \to x) \wedge (x \to y) = (y \to x)$.

Exercise 12 In a linear residuated lattice, for all elements x, y, it holds that $x \leq y$ or $y \leq x$. Assume $x \leq y$. Then $x \vee y = y$ and $x \wedge y = x$ and $x \leftrightarrow y = (x \to y) \wedge (y \to x) = \mathbf{1} \wedge (y \to x) = y \to x = (x \vee y) \to (x \wedge y)$. The case $y \leq x$ is symmetric.

Exercise 13 This follows from Exercise 2 (E), since $\odot$ is commutative.

Exercise 14 $(y \to z) \odot y \leq z$, therefore $[x \odot (y \to z)] \odot y \leq x \odot z$, hence $x \odot (y \to z) \leq y \to (x \odot z)$.

Exercise 15 Consider Gödel structure.
(a): Let $x = 0$. Then, for any $y \in (0,1]$, $y \to 0 = 0$. Hence, $\bigvee_{y \in (0,1]}(y \to 0) = 0$. On the other hand, $\bigwedge_{y \in (0,1]} y = 0$, hence $\bigwedge_{y \in (0,1]} y \to 0 = 0 \to 0 = 1$. Thus,

$$\bigvee_{y \in (0,1]}(y \to 0) < \bigwedge_{y \in (0,1]} y \to 0.$$

(b): Let $x = \frac{1}{2}$. Then, for any $y \in [0, \frac{1}{2})$, $\frac{1}{2} \to y = y$, thus $\bigvee_{y \in [0,\frac{1}{2})}(\frac{1}{2} \to y) = \bigvee_{y \in [0,\frac{1}{2})} y = \frac{1}{2}$, while $\frac{1}{2} \to \bigvee_{y \in [0,\frac{1}{2})} y = 1$. Therefore,

$$\bigvee_{y \in [0,\frac{1}{2})}(\tfrac{1}{2} \to y) < \tfrac{1}{2} \to \bigvee_{y \in [0,\frac{1}{2})} y.$$

Exercise 16 The product operation $\odot$ of a unit interval valued residuated lattice is isotone, commutative and $x \odot \mathbf{1} = x$ holds. Therefore (i), (ii) and (iii), respectively, hold. Moreover, $\odot$ is associative and $x \odot \mathbf{0} = \mathbf{0}$ holds. The converse is true, too; any T-norm can be viewed as a unit interval residuated lattice product operation, where the residuum is defined via (1.29).

Exercise 17 Since the operation $\to$ is isotone in the second variable, we have, for any $i \in \Gamma$, $x \to \bigwedge_{i \in \Gamma} y_i \leq x \to y_i$. Therefore $x \to \bigwedge_{i \in \Gamma} y_i \leq \bigwedge_{i \in \Gamma}(x \to y_i)$. Conversely, for any $i \in \Gamma$, $\bigwedge_{i \in \Gamma}(x \to y_i) \leq x \to y_i$, thus $[\bigwedge_{i \in \Gamma}(x \to y_i)] \odot x \leq y_i$, which implies $[\bigwedge_{i \in \Gamma}(x \to y_i)] \odot x \leq \bigwedge_{i \in \Gamma} y_i$. Therefore $\bigwedge_{i \in \Gamma}(x \to y_i) \leq x \to \bigwedge_{i \in \Gamma} y_i$. We conclude that (1.64) holds.

Since the operation $\rightarrow$ is antitone in the first variable, we have, for each $i \in \Gamma$, $\bigvee_{i\in\Gamma} y_i \rightarrow x \leq y_i \rightarrow x$, which implies $\bigvee_{i\in\Gamma} y_i \rightarrow x \leq \bigwedge_{i\in\Gamma}(y_i \rightarrow x)$. Conversely, as for any $i \in \Gamma$, $\bigwedge_{i\in\Gamma}(y_i \rightarrow x) \leq y_i \rightarrow x$, we have, for any $i \in \Gamma$, $y_i \leq [\bigwedge_{i\in\Gamma}(y_i \rightarrow x)] \rightarrow x$ Thus, $\bigvee_{i\in\Gamma} y_i \leq [\bigwedge_{i\in\Gamma}(y_i \rightarrow x)] \rightarrow x$, or equivalently, $\bigwedge_{i\in\Gamma}(y_i \rightarrow x) \leq \bigvee_{i\in\Gamma} y_i \rightarrow x$. Therefore (1.65) holds.

Exercise 18 By (1.54), $x \rightarrow y \leq y^* \rightarrow x^*$ holds for all elements x, y of a residuated lattice L. Conversely, $y^* \rightarrow x^* \leq x \rightarrow y$ iff $(y^* \rightarrow x^*) \odot x \leq y$ iff $(y^* \rightarrow x^*) \odot (x^* \rightarrow \mathbf{0}) \leq y$, which holds true by (1.45) and the assumption $y^{**} = y$.

Exercise 19 In a residuated lattice L we have, for all $x, y \in L$, $x, y \leq x \vee y$, thus $(x \vee y)^* \leq x^* \wedge y^*$, and $(x^* \wedge y^*) \odot (x \vee y) = [(x^* \wedge y^*) \odot x] \vee [(x^* \wedge y^*) \odot y]$ $\leq (x^* \odot x) \vee (y^*) \odot y) = \mathbf{0}$, hence $x^* \wedge y^* \leq (x \vee y)^*$. We therefore have $x^* \wedge y^* = (x \vee y)^*$.

Exercise 20 The claim follows from the fact $x \odot \bigwedge_{i\in\Gamma} y_i \leq x \odot y_i$, which holds true for all elements x and all subsets $\{y_i\}_{i\in\Gamma}$ of a complete residuated lattice.

Exercise 21 $(x \rightarrow y) \odot z \leq x \rightarrow (y \odot z)$ iff $x \rightarrow y \leq z \rightarrow [x \rightarrow (y \odot z)]$ iff $x \rightarrow y \leq (x \odot z) \rightarrow (y \odot z)$, which holds true by Exercise 2 (F').

5.4 Exercises from Section 1.4

Exercise 1 Since the residuated lattice generated by a continuous T-norm t is linear, it suffices to prove, for real numbers $x, y \in [0, 1]$, that $\min\{x, y\} = t(x, y \rightarrow_t y)$. If $x \leq y$ then $x \rightarrow_t y = 1$ and $\min\{x, y\} = x$ $= t(x, x \rightarrow_t y)$. If $x > y$ then consider a function $t(x, z) : [0, 1] \searrow [0, 1]$, which clearly is continuous, and $t(x, 0) = 0$, $t(x, 1) = x > y$. Thus, there exists $z_0 \in [0, 1]$ such that $t(x, z_0) = y$. By the Galois correspondence, $t(x, x \rightarrow_t y) \leq y$. On the other hand,

$$\begin{aligned} t(x, x \rightarrow_t y) &= t(x, \bigvee\{z \mid t(x, z) \leq y\}) \\ &= \bigvee t(x, \{z \mid t(x, z) \leq y\}) \\ &= t(x, z_0) \\ &= y \\ &= \min\{x, y\}. \end{aligned}$$

Since Łukasiewicz, Gödel and Product multiplication are all continuous T-norms, the corresponding residuated lattices are BL-algebras.

Exercise 2 (a) The only non-trivial thing to show is the Galois correspondence. If $x + y \leq 1$ then $x \leq 1 - y$ and

$$x \odot y = 0 \leq z \text{ iff } x \leq \begin{cases} y \to z = 1 & \text{if } y \leq z \\ \max\{1 - y \vee z\} & \text{if } y > z \end{cases}$$

If $x + y > 1$ then $x > 1 - y$ and

$$x \odot y = \min\{x, y\} \leq z \text{ iff } x \leq \begin{cases} y \to z = 1 & \text{if } y \leq z \\ \max\{1 - y \vee z\} & \text{if } y > z \end{cases}$$

We obtain a residuated lattice.
(b) Take $0 < y < x, x + y < 1$. Then $y < 1 - x$ and $x \wedge y = y \neq 0$, while $x \odot (x \to y) = x \odot (1 - x) = 0$. Thus, (1.70) does not hold, and consequently we do not obtain a BL-algebra.

Exercise 3 If follows by isotonicity of a T-norm t that $t(x, z) \leq t(x, 1) = x$, $t(x, z) \leq t(1, z) = z$. Therefore $t(x, z) \leq x \wedge z$ for any $x, z \in [0, 1]$. Hence $x \Rightarrow y = \bigvee\{z \mid x \wedge z \leq y\} \leq \bigvee\{z \mid t(x, z) \leq y\} = x \to_t y$. Thus, for all $x, y \in [0, 1]$, $x \Leftrightarrow y \leq x \leftrightarrow_t y$.

Exercise 4 Since, for all $a, b \in L$, $(a \to b) \vee (b \to a) = \mathbf{1}$ we have, for any prime *ds* D, that either $a \to b \in D$ or $b \to a \in D$. Conversely, assume either $a \to b \in D$ or $b \to a \in D$ holds for all $a, b \in L$ and D is a *ds*. Let $a \vee b \in D$ and, say, $a \to b \in D$. Now $a \vee b \leq (a \to b) \to b \in D$, therefore $b \in D$. Thus D is prime. To show that L is linear iff each *ds* D of L is prime, first assume L is linear. Then, for all $a, b \in L$, $a \vee b = a$ or $a \vee b = b$. Thus $a \vee b \in D$ iff $a \in D$ or $b \in D$ and so each *ds* D of L is prime. If conversely each *ds* D of L is prime then, in particular, $\{\mathbf{1}\}$ is prime and as, for all $a, b \in L$, $(a \to b) \vee (b \to a) = \mathbf{1}$, either $a \to b = \mathbf{1}$ or $b \to a = \mathbf{1}$, that is either $a \leq b$ or $b \leq a$, whence L is linear.

Exercise 5 In Section 1.2. we show that the proper lattice filters of $[0, 1]$ are the closed intervals $[a, 1]$, $0 < a \leq 1$, and the half open intervals $(a, 1]$, $0 \leq a \leq 1$. All proper deductive systems of $[0, 1]$ are prime, thanks to the linearity of $[0, 1]$.
(a) The only proper *ds* of the Łukasiewicz structure is $\{1\}$. It is also maximal.
(b) On the Gödel structure filters and deductive systems coincide as was proved in Proposition 23
(c) The only proper and therefore also maximal *ds* of the Product structure is the half open interval $\{x \mid 0 < x \leq 1\}$.

Exercise 6 Assume P is a prime *ds* and D, E two such proper *ds* containing P that $E \not\subseteq D$, $D \not\subseteq E$. Then there are two disjoint elements $a, b \in L$ such that $a \in E$, $a \notin D$, $b \in D$, $b \notin E$. Since P is prime either $a \to b \in P$ or $b \to a \in P$. If $a \to b \in P \subseteq E$, then $b \in E$, a contradiction.

Similarly, if $b \to a \in P \subseteq D$, then $a \in D$, another contradiction. Thus, either $E \subseteq D$ or $D \subseteq E$.

Exercise 7 Assume $D \neq \emptyset$ is subset of an BL-algebra L and (i) if $a, b \in D$ then $a \odot b \in D$, (ii) if $a \in D, a \leq b$ then $b \in D$. In such a case there is an element $x \in D \subseteq L$ and as $x \leq \mathbf{1}$ we have $\mathbf{1} \in D$. Assume $x, x \to y \in D$. Then $x \odot (x \to y) \leq y \in D$ and so D is a *ds*. Let conversely D be a *ds*. Assume $a, b \in D$. Since $a \to [b \to (a \odot b)] = \mathbf{1} \in D$ we realize $b \to (a \odot b) \in D$ and, moreover, $a \odot b \in D$. Thus (i) holds. To verify (i) let $a \in D$, $a \leq b$. Then $a \to b = \mathbf{1} \in D$, whence $b \in D$.

Exercise 8 $\mathbf{1} \sim \mathbf{1}$ so $\mathbf{1} \in D$. Assume $x, y \in D$. Then $x \sim \mathbf{1}$, $y \sim \mathbf{1}$ and therefore $(x \odot y) \sim (\mathbf{1} \odot \mathbf{1}) = \mathbf{1}$, whence $x \odot y \in D$. Let $x \in D$, $x \leq y$. Then $x \to y = \mathbf{1} \in D$ and $x \sim \mathbf{1}$, $y \sim y$ so $\mathbf{1} = x \to y \sim \mathbf{1} \to y = y$, i.e. $y \sim \mathbf{1}$. By Exercise 7, D is a *ds* of D.

Exercise 9 We have seen that $x \sim_D y$ iff $(x \to y) \odot (y \to x) \in D$, where D is a *ds* of a BL-algebra L, is a congruence relation with respect to $\odot$ and $\to$. Since the operations $\wedge$ and $\vee$ can be expressed by means of $\odot$ and $\to$ we conclude $\sim_D$ is a congruence with respect to $\wedge$ and $\vee$, too.

Exercise 10 Notice first that if $\equiv$ is a congruence and $\mathbf{0} \not\equiv \mathbf{1}$ then $\equiv$ is proper and, if $\mathbf{0} \equiv \mathbf{1}$ then $\mathbf{0} = (\mathbf{0} \odot x) \equiv (\mathbf{1} \odot x) = x$ whence, for all $x, y \in L$, $x \equiv y$ and so $\equiv$ is not proper. Therefore a proper congruence generates a proper deductive system. Also the converse holds. Indeed, if D is a proper *ds* then $\mathbf{0} \notin D$, whence $(\mathbf{1} \to \mathbf{0}) \odot (\mathbf{0} \to \mathbf{1}) = \mathbf{0} \odot \mathbf{1} = \mathbf{0} \notin D$, thus $\mathbf{0} \not\sim_D \mathbf{1}$. Assume now D is a maximal *ds* of L but $\sim_D$ is not a maximal congruence relations on L. Then there is a proper congruence $\equiv$ such that, for some $x, y \in L$, $x \equiv y$, $x \not\sim_D y$ while, for any $a, b \in L$, $a \sim_D b$ implies $a \equiv b$. If both $x \to y \in D$ and $y \to x \in D$ then $(x \to y) \odot (y \to x) \in D$ implying a contradiction $x \sim_D y$. Assume therefore, say, $x \to y \notin D$. Now $\equiv$, being a proper congruence, generates a proper *ds* $D_\equiv$. Then $D \subseteq D_\equiv$. Indeed, if $a \in D$ then $a \sim_D \mathbf{1}$, thus $a \equiv \mathbf{1}$ and so $a \in D_\equiv$. Since $x \equiv y$, $\mathbf{1} = x \to x \equiv x \to y$. Therefore $(x \to y) \in D_\equiv$ and $(x \to y) \neq \mathbf{0}$ as $D_\equiv$ is proper. This contradicts the fact D is a maximal *ds*. Consequently, $\sim_D$ is a maximal congruence. To prove the converse, assume $\sim$ is a maximal congruence but $D_\sim$ is not a maximal *ds*. Then there is a proper *ds* E such that $D_\sim \subseteq E$ and an element $x \in L$ with $x \in E$, $x \notin D_\sim$. Now E defines a proper congruence $\sim_E$ and if $a \sim b$, then $(a \to b) \odot (b \to a) \in D_\sim$, thus $(a \to b) \odot (b \to a) \in E$, whence $a \sim_E b$. Moreover, $x \sim_E \mathbf{1}$, $x \not\sim \mathbf{1}$. But this contradicts the fact $\sim$ is a maximal congruence. Therefore $D_\sim$ is a maximal *ds*.

Exercise 11 (1.84): $h(x^*) = h(x \to \mathbf{0}_L) = h(x) \to h(\mathbf{0}_L) = h(x) \to \mathbf{0}_K = h(x)^*$, and $h(\mathbf{1}_L) = h(\mathbf{0}_L \to \mathbf{0}_L) = h(\mathbf{0}_L) \to h(\mathbf{0}_L) = \mathbf{0}_K \to \mathbf{0}_K = \mathbf{1}_K$.
(1.85): If $x \leq y$ then $x \to y = \mathbf{1}_L$, thus $h(x \to y) = h(x) \to h(y) = \mathbf{1}_K$, hence $h(x) \leq h(y)$.
(1.86): follows by the fact that $\wedge$ and $\vee$ are definable by $\odot$ and $\to$.
(1.87): $\mathbf{1}_K = h(\mathbf{1}_L) \in h(D)$, if $h(a), h(a) \to h(b) \in h(D)$ then $h(a \to b) \in h(D)$, thus $a, a \to b \in D$, hence $b \in D$, whence $h(b) \in h(D)$.

Exercise 12 Let $m \geq 1$. The operation $\odot$ on $S(m)$ is clearly commutative, isotone and $\frac{p}{m} \odot 1 = \frac{p}{m}$ for all $0 \leq p \leq m$. It is also associative as, for all $0 \leq p, k, s \leq m$,

$$\begin{aligned}
\frac{p}{m} \odot (\frac{k}{m} \odot \frac{s}{m}) &= \frac{p}{m} \odot \frac{\max\{0, k+s-m\}}{m} \\
&= \frac{\max\{0, p + \max\{0, k+s-m\} - m\}}{m} \\
&= \frac{\max\{0, \max\{0, p+k-m\} + s - m\}}{m} \\
&= \frac{\max\{0, p+k-m\}}{m} \odot \frac{s}{m} \\
&= (\frac{p}{m} \odot \frac{k}{m}) \odot \frac{s}{m}.
\end{aligned}$$

To verify the Galois correspondence we reason

$$\tfrac{p}{m} \odot \tfrac{k}{m} = \tfrac{\max\{0,p+k-m\}}{m} \leq \tfrac{s}{m} \text{ iff } \tfrac{p}{m} \leq \tfrac{\min\{m,m-k+s\}}{m} \leq \tfrac{s}{m} = \tfrac{k}{m} \to \tfrac{s}{m}.$$

We have shown that $S(m)$ is a residuated lattice. It is a BL-algebra, too. Indeed, as $S(m)$ is linear we need to show only (1.70). Let $p \leq k$. Then

$$\begin{aligned}
\frac{p}{m} &= \min\{\frac{p}{m}, \frac{k}{m}\} = \frac{\max\{0, p + \min\{m, m-p+k\} - m\}}{m} \\
&= \frac{p}{m} \odot \frac{\min\{m, m-p+k\}}{m} = \frac{p}{m} \odot (\frac{p}{m} \to \frac{k}{m}).
\end{aligned}$$

$S(m)$ is locally finite as, for all $0 < p < m$, $(\frac{p}{m})^m = 0$.

Exercise 13 $x \in^{\perp} A$ iff $\forall a \in A : x \vee a = 1$ iff $\forall a \in A : x \to a = a$ and $a \to x = x$ iff $\forall a \in A : x \in^{\perp}\{a\}$ iff $x \in \bigcap_{a \in A} {}^{\perp}\{a\} = \bigcap_{a \in A} D^a$.

Exercise 14 Assume $X \subseteq Y$. ${}^{\perp}Y = \{a \in L \mid a \vee y = \mathbf{1} \text{ for all } y \in Y\}$. If $a \in^{\perp} Y$ then $a \vee x = \mathbf{1}$ for all $x \in X \subseteq Y$, hence $a \in^{\perp} X$ and therefore ${}^{\perp}Y \subseteq {}^{\perp}X$.

5.5 Exercises from Section 2.1

Exercise 1 Let $x \leq y$. Then $x \to y = \mathbf{1}$.

(a): By (2.2), it holds that $x \to y \leq (y \to z) \to (x \to z)$. Thus $\mathbf{1} =$ $(y \to z) \to (x \to z)$, therefore $(y \to z) \leq (x \to z)$, i.e. $\to$ is antitone in the first varible.
(b): By (2.12), $x \to y \leq (z \to x) \to (z \to y)$, thus $(z \to x) \leq (z \to y)$ which proves $\to$ is isotone in the second variable.

Exercise 2 Since $(x \to \mathbf{1}^*) \to (x \to \mathbf{1}^*) = \mathbf{1}$ we have, by (2.13), that $x \to [(x \to \mathbf{1}^*) \to \mathbf{1}^*] = \mathbf{1}$. Thus $x \leq x^{**}$. By (2.4), $(x^* \to \mathbf{1}^*) \to (\mathbf{1} \to x)$ $= \mathbf{1}$, i.e. $(x^{**}) \to x = \mathbf{1}$. Hence $x^{**} \leq x$. We conclude that (2.20) holds. By (2.4), (2.20) and (2.4), $x^* \to y^* \leq y \to x = y^{**} \to x^{**} \leq x^* \to y^*$. Therefore (2.21) holds. If $x \leq y$, then $x \to y = \mathbf{1}$ thus, by (2.20), $x^{**} \to y^{**}$ $= \mathbf{1}$ hence, by (2.4), $\mathbf{1} \to (y^* \to x^*) = \mathbf{1}$, which implies $y^* \leq x^*$. By a similar argument, $y^* \leq x^*$ implies $x^{**} \leq y^{**}$ so, by (2.20), $x \leq y$. Therefore (2.22) holds.

Exercise 3 By (2.24) and (2.20), $(x \wedge y)^* = (x^* \vee y^*)^{**} = x^* \vee y^*$, and similarly, $(x \vee y)^* = (x^{**} \vee y^{**})^* = x^* \wedge y^*$.

Exercise 4 (a): By definition, (2.15) and (2.20), $x \odot \mathbf{1} = (x \to \mathbf{1}^*)^* =$ $x^{**} = x$.
(b): By definition, (2.22), (2.20), (2.5), (2.13), (2.5), (2.21), $x \odot y = (x \to$ $y^*)^* \leq z$ iff $z^* \leq x \to y^*$ iff $\mathbf{1} = z^* \to (x \to y^*)$ iff $\mathbf{1} = x \to (z^* \to y^*)$ iff $x \leq z^* \to y^*$ iff $x \leq y \to z$.

Exercise 5 (2.35) holds in a Wajsberg algebra as

$$
\begin{array}{rcll}
(x \wedge y) \to z & = & z^* \to (x \wedge y)^* & \text{(by (2.21))} \\
& = & z^* \to (x^* \vee y^*) & \text{(by (2.25))} \\
& = & z^* \to [(x^* \to y^*) \to y^*] & \text{(by (2.23))} \\
& = & z^* \to [(y^* \to x^*) \to x^*] & \text{(by (2.3))} \\
& = & (y^* \to x^*) \to (z^* \to x^*) & \text{(by (2.13))} \\
& = & (x \to y) \to (x \to z) & \text{(by (2.21)).}
\end{array}
$$

Exercise 6 Consider first the right column if equations (2.39) - (2.48). (2.39), (2.40), (2.43), (2.46), (2.47) and (2.48) follow immediately from (2.27), (2.28), (2.30), (1.10), (1.12) and (1.28), respectively.
(2.41): $x \odot x^* \leq \mathbf{0}$ iff $x \leq x^* \to \mathbf{0} = x^{**}$, which holds true by (2.20). (2.42): $x \odot \mathbf{0} \leq \mathbf{0}$ iff $x \leq \mathbf{0} \to \mathbf{0} = \mathbf{1}$, which holds true.
(2.44): By (2.26), (2.20), (2.36), we have $x \odot y = (x \to y^*)^* = (x^{**} \to y^*)^*$ $= (x^* \oplus y^*)^*$. Therefore $(x \odot y)^* = (x^* \oplus y^*)^{**} = x^* \oplus y^*$.
(2.45): $\mathbf{1} = \mathbf{0} \to \mathbf{0} = \mathbf{0}^*$, therefore $\mathbf{1}^* = \mathbf{0}$. Now consider the left column of equations (2.39) - (2.48). (2.45), (2.46) and (2.47) follow immediately (2.20), (1.10) and (1.12), respectively.
(2.44): By (2.20) and the other part of (2.44), we have $x \oplus y = x^{**} \oplus y^{**}$ $= (x^* \odot y^*)^*$. Therefore $(x \oplus y)^* = x^* \odot y^*$.

(2.39): $x \oplus y = (x^* \odot y^*)^* = (y^* \odot x^*)^* = y \oplus x$.(2.40): $x \oplus (y \oplus z) = [x^* \odot (y \oplus z)^*]^* = [x^* \odot (y^* \odot z^*)]^* = [(x^* \odot y^*) \odot z^*]^* = [(x \oplus y)^* \odot z^*]^* = (x \oplus y) \oplus z$.(2.41): $x \oplus x^* = (x^* \odot x)^* = \mathbf{0}^* = \mathbf{1}$.
(2.42): $x \oplus \mathbf{1} = (x^* \odot \mathbf{1}^*)^* = (x^* \odot \mathbf{0})^* = \mathbf{0}^* = \mathbf{1}$.
(2.43): $x \oplus \mathbf{0} = (x^* \odot \mathbf{0}^*)^* = (x^* \odot \mathbf{1})^* = x^{**} = x$.
(2.48): $x \oplus (y \wedge z) = [x^* \odot (y \wedge z)^*]^* = [x^* \odot (y^* \vee z^*)]^* = [(x^* \odot y^*) \vee (x^* \odot z^*)]^* = (x^* \odot y^*)^* \wedge (x^* \odot z^*)^* = (x \oplus y) \wedge (x \odot z)$.

Exercise 7 (a): $(x \wedge y)^* = [(x \oplus y^*) \odot y]^* = (x \oplus y^*)^* \oplus y^* = (x^* \odot y^{**}) \oplus y^* = x^* \vee y^*$.
(b): $x \vee (x \wedge y) = (y \wedge x) \vee x = [(y \oplus x^*) \odot x] \vee x = \{[(y \oplus x^*) \odot x] \odot x^*\} \oplus x = [(y \oplus x^*) \odot (x \odot x^*)] \oplus x = [(y \oplus x^*) \odot \mathbf{0}] \oplus x = \mathbf{0} \oplus x = x$.
(c): If $x \odot y = \mathbf{1}$, then $x^* \oplus y^* = \mathbf{0}$, thus $x^* = y^* = \mathbf{0}$, hence $x = y = \mathbf{1}$.
(d): If $x \wedge y = \mathbf{1}$, then $x^* \vee y^* = \mathbf{0}$, thus $x^* = y^* = \mathbf{0}$, hence $x = y = \mathbf{1}$.
(e): If $x \leq y$, then $x \wedge y = x$ and therefore $(x \wedge z) \wedge (y \wedge z) = (x \wedge y) \wedge (z \wedge z) = x \wedge z$, thus $x \wedge z \leq y \wedge z$.
(f): If $x, y \leq z$, then $z = x \vee z = y \vee z$ and $z = z \vee z = (x \vee z) \vee (y \vee z) = (x \vee y) \vee z$. Therefore $x \vee y \leq z$.
(g): If $x \leq y$, then $x \vee y = y$ and $(x \odot z) \vee (y \odot z) = z \odot (y \vee x) = z \odot y = y \odot z$. Therefore $x \odot z \leq y \odot z$.

Exercise 8 If (2.79) holds in a residuated lattice L then, for all $x, y \in L$, $x^{**} = x$ and, by Exercise 1.3.18, $x \to y = (y^* \to x^*)$. Thus

$$[(x^* \to y^*) \to y^*]^* = [\text{l.u.b.}\{x^*, y^*\}]^* = \text{g.l.b.}\{x, y\},$$

where the last equation follows from Exercise 1.3.19. But $[(x^* \to y^*) \to y^*]^* = [(y \to x) \to y^*]^* = [(y \to x) \odot y]^{**} = (y \to x) \odot y$. By a similar argument we see that $\text{g.l.b}\{x, y\} = x \odot (x \to y)$.

Exercise 9 Boolean algebras are distributive, residuated lattices where l-complement coincides with complement and $x^{**} = x$ holds. Moreover, Boolean algebras are soft algebras. Thus $(x \to y) \to y = (x^* \vee y)^* \vee y = (x^{**} \wedge y^*) \vee y = (x \wedge y^*) \vee y = (x \vee y) \wedge (y^* \vee y) = (x \vee y) \wedge \mathbf{1} = x \vee y$. Since $x \vee y = y \vee x$, we conclude that $(x \to y) \to y = (y \to x) \to x$. Hence, Boolean algebras are Wajsberg algebras.

Exercise 10 Assume $x \odot x = x$. Then $x \oplus x = (x^* \odot x^*)^* = x^{**} = x$. Conversely, let $x \oplus x = x$. Then $x \odot x = (x^* \oplus x^*)^* = x^{**} = x$.

Exercise 11 For any $p \geq 1$, $x, y \in [0, 1]$,

$$\begin{aligned}
(x \to y) \to y &= \min\{1, \sqrt[p]{1 - [\,\min\{1, \sqrt[p]{1 - x^p + y^p}\}]^p + y^p}\} \\
&= \min\{1, \sqrt[p]{1 - \ \min\{1, 1 - x^p + y^p\} + y^p}\} \\
&= \min\{1, \sqrt[p]{1 - 1 + y^p}\} \\
&= y \ \text{ iff } \ 1 \leq 1 - x^p + y^p \ \text{ iff } \ x \leq y.
\end{aligned}$$

in any generalized Łukasiewicz structure, and

$$\begin{aligned}(x \to y) \to y &= \min\{1, \sqrt[p]{1 - 1 + x^p - y^p + y^p}\} \\ &= x \text{ iff } 1 - x^p + y^p \leq 1 \text{ iff } y \leq x.\end{aligned}$$

Hence $(x \to y) \to y = \max\{x, y\}$, whence generalized Łukasiewicz structures define MV-algebra structures. In the other structures,

$$(x \to y) \to y = \min\{1, \sqrt[p]{[1 - \min\{1, \sqrt[p]{(1-x)^p + y^p}\}]^p + y^p}\}$$

By direct computation we see that $(0.9 \to 0.1) \to 0.1 = 0.9$ iff $p = 1$. Thus $(x \to y) \to y = \max\{x, y\}$ does not generally hold. Therefore these structures do not generate, in general, a Wajsberg algebra structure.

Exercise 12 By (2.25) and (2.20), (iii) and (iv) are equivalent. Assume (ii). Then $x^* \vee y = (y \odot x^{**}) \oplus x^* = (y \odot x) \oplus x^* = x \oplus x^* = \mathbf{1}$. Thus (iii) holds. Now assume (iii). Then $x = x \odot \mathbf{1} = x \odot (x^* \vee y) = (x \odot x^*) \vee (x \odot y) = \mathbf{0} \vee (x \odot y) = x \odot y$, so (ii) and (iii) are equivalent. The connection between (iv) and (i) is symmetric: if (iv) holds then $y = y \oplus \mathbf{0} = y \oplus (x \wedge y^*) = (y \oplus x) \wedge (y \oplus y^*) = (y \oplus x) \wedge \mathbf{1} = y \oplus x$ and, if (i) holds, then $x \wedge y^* = y^* \odot (y \oplus x) = y^* \odot y = \mathbf{0}$.

Exercise 13 First we realize that f^{-1} is continuous, strictly increasing, $f^{-1}(0) = 0$, $f^{-1}(k) = 1$ and $f^{-1}(f(x)) = x$ for any $x \in I$. Let $x, y \in I$. Then $x \leq y$ iff $f(x) \leq f(y)$ iff $k - f(y) \leq k - f(x)$ iff $f^{-1}(k - f(y)) \leq f^{-1}(k - f(x))$ iff $y^* \leq x^*$. Moreover, $k - f(x) + f(y) \in [0, k]$ iff $k - f(x) + f(y) \leq k$ iff $f(y) \leq f(x)$ iff $y \leq x$. Now define a binary operation $\to$ on I, for any $x, y \in I$, via $x \to y = f^{-1}(k - f(x) + f(y))$ if $y < x$ and 1 otherwise. We show that $\langle [0,1], \to, ^*, 1\rangle$ is a Wajsberg algebra by verifying the axioms (2.1) - (2.4). So, let $x, y, z \in I$.
(2.1): $1 \to x = f^{-1}(k - f(1) + f(x)) = f^{-1}(f(x)) = x$.
(2.2): $y \to z = f^{-1}(k - f(y) + f(z))$, $x \to z = f^{-1}(k - f(x) + f(z))$, thus $(y \to z) \to (x \to z) = f^{-1}(k - k + f(y) - f(z) + k - f(x) + f(z)) = f^{-1}(k - f(x) + f(y)) = x \to y$. Thus $(x \to y) \to [(y \to z) \to (x \to z)] = 1$.
(2.3): If $x \leq y$, then $(x \to y) \to y = 1 \to y = y = \max\{x, y\}$. If $y < x$, then $(x \to y) \to y = f^{-1}(k - k + f(x) - f(y) + f(y)) = x = \max\{x, y\}$. Therefore $(x \to y) \to y = \max\{x, y\} = (y \to x) \to x$.
(2.4): Assume $y \leq x$. Then $(x^* \to y^*) \to (y \to x) = (x^* \to y^*) \to 1 = 1$. If $x < y$, then $y^* < x^*$ and $x^* \to y^* = f^{-1}(k - k + f(x) + k - f(y)) = f^{-1}(k - f(y) + f(x)) = y \to x$. Therefore $(x^* \to y^*) \to (y \to x) = 1$.

Exercise 14 Clearly $\mathbf{0}^*, \mathbf{1}^* \in MV(L)$. Since $x^* \to y^* = x^* \to (y \to \mathbf{0}) = (x^* \odot y)^* \in MV(L)$, $MV(L)$ is closed with respect to residuum. It is obvious that (2.1) and (2.2) hold in $MV(L)$. To prove (2.4) we reason $[(x^{**} \to y^{**}) \odot x] \odot y^* \leq [(x^{**} \to y^{**}) \odot x^{**}] \odot y^* \leq y^{**} \odot y^* = \mathbf{0}$, therefore $(x^{**} \to y^{**}) \odot y^* \leq x^*$, whence $(x^{**} \to y^{**}) \leq y^* \to x^*$, and so (2.4) holds. To prove (2.3), we will need the following results

$$x^* \to y^* = y \to x^{**} \text{ and } x^{***} = x^*,$$

which hold in all residuated lattices. We also utilize a BL-algebra property $x^{**} \odot (x^{**} \to y) = y \odot (y \to x^{**})$. Now we are ready to reason

$$\begin{aligned}
(x^* \to y^*) \to y^* &= (y \to x^{**}) \to y^* \\
&= [y \odot (y \to x^{**})]^* \\
&= [x^{**} \odot (x^{**} \to y)]^* \\
&= (x^{**} \to y) \to x^{*(**)} \\
&\geq (x^{**} \to y^{**}) \to x^* \\
&= (y^* \to x^{*(**)}) \to x^*.
\end{aligned}$$

By a symmetric argument $(x^* \to y^*) \to y^* \leq (y^* \to x^*) \to x^*$. Therefore (2.3) holds.

5.6 Exercises from Section 2.2

Exercise 1 In a complete MV-algebra L,

$$(\bigvee_{i\in\Gamma} y_i^*)^* = \bigwedge_{i\in\Gamma} y_i^{**} = \bigwedge_{i\in\Gamma} y_i.$$

Therefore $\bigvee_{i\in\Gamma} y_i^* = (\bigwedge_{i\in\Gamma} y_i)^*$.

Exercise 2 The operation $\wedge$ on a complete MV-algebra L is associative, commutative, isotone and $a \wedge \mathbf{1} = a$ holds for all elements $a \in L$, moreover, by defining the residuum $\Rightarrow$ of $\wedge$ via (1.28) we obtain an adjoin couple $\langle \wedge, \Rightarrow \rangle$ and the claim now follows from (1.27).

Exercise 3 In a complete MV-algebra L,

$$x \to \bigvee_{i\in\Gamma} y_i = x^* \oplus \bigvee_{i\in\Gamma} y_i = \bigvee_{i\in\Gamma}(x^* \oplus y_i) = \bigvee_{i\in\Gamma}(x \to y_i)$$

and

$$\bigwedge_{i\in\Gamma} y_i \to x = (\bigwedge_{i\in\Gamma} y_i)^* \oplus x = \bigvee_{i\in\Gamma} y_i^* \oplus x = \bigvee_{i\in\Gamma}(y_i^* \oplus x) = \bigvee_{i\in\Gamma}(y_i \to x).$$

Therefore L is continuous.

Exercise 4 Let $a \in [0,1]$. Since g_a is increasing we have, by (2.93), $\lim_{x\to b^-} g_a(x) = \bigvee_{x<b}(a \to x) = a \to \bigvee_{x<b} x = a \to b = f_a(b)$. By (1.64), $\lim_{x\to b^+} g_a(x) = \bigwedge_{b<x}(a \to x) = a \to \bigwedge_{b<x} x = a \to b = g_a(b)$. Thus, for each $b \neq 0, 1$, $\lim_{x\to b} g_a(x) = g_a(b)$ and $\lim_{x\to 1^-} g_a(x) = g_a(1)$, $\lim_{x\to 0^+} g_a(x) = g_a(0)$. We conclude that g_a is continuous in the unit interval. Assume the residuated lattice on the unit interval under consideration

is complete but some g_a is not a continuous function. Then g_a is not continuous at some point $b \in [0,1]$. Since $\lim_{x\to b^+} g_a(x) = g_a(b)$, we have $b \neq 0$ and $\lim_{x\to b^-} g_a(x) \neq g_a(b)$. Thus, by (1.67), $\lim_{x\to b^-} g_a(x) = \bigvee_{x<b}(a \to x) < a \to \bigvee_{b<x} x = a \to b = f_a(b)$. Therefore (2.93) does not hold.

Exercise 5 Let $x, y \in [0,1]$. Then $x \oplus y = \min\{\ln(e^x + e^y - 1), 1\}$, $x^* = \ln(e + 1 - e^x)$, $x \to y = \ln(e - e^x + e^y)$ if $x > y$ and 1 otherwise. $x \odot y = \min\{\ln(e^x + e^y - e), 0\}$.

Exercise 6 By (2.92) and (1.28), $x \odot \bigwedge_{i\in\Gamma} y_i = \bigwedge_{i\in\Gamma}(x \odot y_i)$ and $x \odot \bigvee_{i\in\Gamma} y_i = \bigvee_{i\in\Gamma}(x \odot y_i)$, respectively, hold in any complete MV-algebra defined on the unit interval. Since functions h_a, $a \in I$, are increasing, it is easy to see that, for each $b \in [0,1]$, $\lim_{x\to b^-} h_a(x) = h_a(b) = \lim_{x\to b^+} h_a(x)$ (considering, of course, only right or left limits if $b = 0$ or $b = 1$, respectively). Thus, the functions h_a are continuous. Similarly, by (2.87) and (2.91), for functions k_a.

Exercise 7 (a): We observe, by using (2.87) in the second equation, that

$$\begin{aligned}
(x \oplus y) \odot [(x \vee y) \oplus (x \wedge y)]^* &= (x \oplus y) \odot (x^* \wedge y^*) \odot (x \wedge y)^* \\
&= [(x \oplus y) \odot x^*] \wedge [(x \oplus y) \odot y^*] \odot (x \wedge y)^* \\
&= [(x^* \wedge y) \wedge (y^* \wedge x)] \odot (x \wedge y)^* \\
&\leq (x \wedge y) \odot (x \wedge y)^* \\
&= \mathbf{0}.
\end{aligned}$$

Therefore $(x \oplus y) \leq (x \vee y) \oplus (x \wedge y)$. Conversely,

$$\begin{aligned}
(x \vee y) \oplus (x \wedge y) &= [x \oplus (x \wedge y)] \vee [y \oplus (x \wedge y)] && \text{by (2.91)} \\
&= [2x \wedge (x \oplus y)] \vee [2y \wedge (x \oplus y)] && \text{by (2.87)} \\
&= (x \oplus y) \wedge (2x \vee 2b) && \text{distributivity} \\
&\leq x \oplus y.
\end{aligned}$$

Therefore (a) holds. (b) is just a dual statement of (a). Indeed,

$$\begin{aligned}
x \odot y &= (x \odot y)^{**} \\
&= (x^* \oplus y^*)^* \\
&= [(x^* \vee y^*) \oplus (x^* \wedge y^*)]^* \\
&= [(x \wedge y)^* \oplus (x \vee y)^*]^* \\
&= [(x \wedge y) \odot (x \vee y)]^{**} \\
&= (x \wedge y) \odot (x \vee y).
\end{aligned}$$

Exercise 8 (a) $(nx)^* = (x^*)^n$ is obviously true when $n = 0$ or 1. Assume (induction hypothesis) it is true for all positive integers $\leq n$. Then $[(n+1)x]^* = (nx \oplus x)^* = (x^*)^n \odot x^* = (x^*)^{n+1}$. Thus, the claim holds for all $n \geq 0$. For (b) , we show only the induction step: $(x^{n+1})^* = (x^n \odot x)^* = (x^n)^* \oplus x^* = n(x^*) \oplus x^* = (n+1)x^*$.

Exercise 9 (a) The claim is obviously true in case $n = 1$. Let us assume (induction hypothesis) it is true for all natural numbers $\leq n$. Then

$$\begin{aligned}(x \vee y)^{n+1} &= (x \vee y)^n \odot (x \vee y) \\ &= (x^n \vee y^n) \odot (x \vee y) \\ &= [x^n \odot (x \vee y)] \vee [y^n \odot (x \vee y)] \\ &= x^{n+1} \vee (x^n \odot y) \vee (y^n \odot x) \vee y^{n+1} \\ &= (x^{n+1} \vee y^{n+1}) \vee [(x^n \odot y) \vee (y^n \odot x)].\end{aligned}$$

But

$$\begin{aligned}(x^n \odot y) \vee (y^n \odot x) &= (x \odot y) \odot (x^{n-1} \vee y^{n-1}) \\ &= (x \vee y) \odot (x \wedge y) \odot (x \vee y)^{n-1} \\ &= (x \wedge y) \odot (x \vee y)^n \\ &= (x \wedge y) \odot (x^n \vee y^n) \\ &= [x^n \odot (x \wedge y)] \vee [y^n \odot (x \wedge y)] \\ &= [x^{n+1} \wedge (x^n \odot y)] \vee [(y^n \odot x) \wedge y^{n+1}] \\ &\leq x^{n+1} \vee y^{n+1}.\end{aligned}$$

Therefore $(x \vee y)^{n+1} = x^{n+1} \vee y^{n+1}$.
(d) can be seen now dually:

$$\begin{aligned}n(x \wedge y) &= [n(x \wedge y)]^{**} \\ &= \{[(x \wedge y)^*]^n\}^* \\ &= [(x^* \vee y^*)^n]^* \\ &= [(x^*)^n \vee (y^*)^n]^* \\ &= [(nx)^* \vee (ny)^*]^* \\ &= nx \wedge ny.\end{aligned}$$

(b): Case of $n = 1$ is obvious. The induction step can be seen as follows:

$$\begin{aligned}(x \wedge y)^{n+1} &= (x \wedge y)^n \odot (x \wedge y) \\ &= (x^n \wedge y^n) \odot (x \wedge y) \\ &= [x^n \odot (x \wedge y)] \wedge [y^n \odot (x \wedge y)] \\ &= x^{n+1} \wedge (x^n \odot y) \wedge (y^n \odot x) \wedge y^{n+1} \\ &= (x^{n+1} \wedge y^{n+1}) \wedge (x^n \odot y) \wedge (y^n \odot x).\end{aligned}$$

Now

$$\begin{aligned}(x^n \odot y) \wedge (y^n \odot x) &= (x \odot y) \odot (x^{n-1} \wedge y^{n-1}) \\ &= (x \wedge y) \odot (x \vee y) \odot (x \wedge y)^{n-1} \\ &= (x \vee y) \odot (x \wedge y)^n \\ &= (x \vee y) \odot (x^n \wedge y^n) \\ &= [x^n \odot (x \vee y)] \vee [y^n \odot (x \vee y)] \\ &= [x^{n+1} \vee (x^n \odot y)] \vee [y^{n+1} \vee (y^n \odot x)] \\ &\geq x^{n+1} \wedge y^{n+1}.\end{aligned}$$

This implies $(x \wedge y)^{n+1} = x^{n+1} \wedge y^{n+1}$.

(c) is again a dual statement to (b). Indeed,

$$\begin{aligned} n(x \vee y) &= [n(x \vee y)]^{**} \\ &= \{[(x \vee y)^*]^n\}^* \\ &= [(x^* \wedge y^*)^n]^* \\ &= [(x^*)^n \wedge (y^*)^n]^* \\ &= [(nx)^* \wedge (ny)^*]^* \\ &= nx \vee ny. \end{aligned}$$

Exercise 10 If $x \wedge y = \mathbf{0}$ then $n(x \wedge y) = \mathbf{0}$ and, by Exercise 9 (d), $nx \wedge ny = \mathbf{0}$ holds for all natural numbers n.

Exercise 11 If $ord(x \oplus y) = n < \infty$ then $(x \oplus y)^n = \mathbf{0}$, whence $[(x \oplus y)^n]^* = \mathbf{1}$, hence $n(x \oplus y)^* = \mathbf{1}$ and, therefore $n(x^* \odot y^*) = \mathbf{1}$. Moreover, $(x \oplus y) \vee (x^* \oplus y^*) = \mathbf{1}$, thus $(x^* \odot y^*) \wedge (x \odot y) = \mathbf{0}$, therefore $n(x^* \odot y^*) \wedge n(x \odot y) = \mathbf{0}$, which implies $n(x^* \odot y^*) = \mathbf{0}$. But $x^* \odot y^* \leq n(x^* \odot y^*)$, thus $x^* \odot y^* = \mathbf{0}$.

Exercise 12 We note first that $x^* \neq \mathbf{1}$ as $x \neq \mathbf{0}$. Suppose $x^* \leq y$. Then $y^* \leq x$ so that either $y^* = \mathbf{0}$ or $y^* = x$, that is either $y = \mathbf{1}$ or $y = x^*$. Therefore x^* is a co-atom.

Exercise 13 Since $\uplus$ is associative it is enough to show that $(m, x) \leq (n, y)$ implies

$$(*) \qquad (k, z) \uplus (m, x) \leq (k, z) \uplus (n, y).$$

By definition,

$$(k, z) \uplus (m, x) = \begin{cases} (k + m, z \oplus x) & \text{if } z \oplus x < \mathbf{1} \\ (k + m + 1, z \odot x) & \text{if } z \oplus x = \mathbf{1} \end{cases}$$

and

$$(k, z) \uplus (n, y) = \begin{cases} (k + n, z \oplus y) & \text{if } z \oplus y < \mathbf{1} \\ (k + n + 1, z \odot y) & \text{if } z \oplus y = \mathbf{1}. \end{cases}$$

We have four cases. Case 1. If $m < n$, then $k + m < k + n$ and, case 2, if $m = n$, $x \leq y$, $z \oplus x < \mathbf{1}$, $z \oplus y < \mathbf{1}$, then $z \oplus x \leq z \oplus y$. In both cases $(k + m, z \oplus x) \leq (k + n, z \oplus y)$ and so $(*)$ holds. Case 3. If $m = n$, $x \leq y$, $z \oplus x < \mathbf{1}$, $z \oplus y = \mathbf{1}$ then $(k + m, z \oplus x) \leq (k + n + 1, z \odot y)$ and again $(*)$ holds. Case 4. If $m = n$, $x \leq y$, $z \oplus x = z \oplus y = \mathbf{1}$, then $z \odot x \leq z \odot y$ thus $(k + m + 1, z \odot x) \leq (k + n + 1, z \odot y)$, hence $(*)$ holds.

Exercise 14 The claim that a mapping $\varphi_M : L \overset{h}{\mapsto} L/M \overset{k}{\mapsto} [0, 1]$ is a BL-homomorphism follows easily by the fact that h and m are BL-homomorphisms, φ_M e.g. preserves the operation $\odot$ as, for all $x, y \in L$, $\varphi_M(x \odot y) = k(|x \odot y|) = k(|x|) \odot k(|y|) = \varphi_M(x) \odot \varphi_M(y)$. The preservation of the other BL-operations and zero and unit can be verified equally.

Exercise 15 (i): $k_1(x) = \underbrace{(x \oplus \ldots \oplus x)^*}_{m-1 \text{ terms}} \leq x$ iff $\underbrace{x \oplus \ldots \oplus x}_{m \text{ terms}} = \mathbf{1}$, which holds true.
(ii): If $k_1(x) = \mathbf{0}$ for some $x \neq \mathbf{0}, \mathbf{1}$ then $(O(x) - 1)x = \mathbf{1}$, a contradiction and, if $k_1(x) = \mathbf{1}$ then $(m-1)x = \mathbf{0}$ which is not the case as $x \neq \mathbf{0}$, $2 \leq m$. Therefore $k_1(x) \neq \mathbf{0}, \mathbf{1}$ for all $x \neq \mathbf{0}, \mathbf{1}$, hence $k_i(x) = k_1(k_{i-1}(x)) \neq \mathbf{0}$ for all $i > 1$.
In Łukasiewicz structure, $k_1(x) = 1 - (m-1) \cdot x$, where $O(x) = m$. Thus, (iii): if $O(x) = O(y) = m$, $x < y$, then $(m-1) \cdot x < (m-1) \cdot y$ and therefore $k_1(y) = 1 - (m-1) \cdot y < 1 - (m-1) \cdot x = k_1(x)$.
(iv) $k_1(x) = 1 - (m-1) \cdot x = 1 - m \cdot x + x = x$ iff $1 = m \cdot x$ iff $x = \frac{1}{m}$ $O(x) = m > 1$.

Exercise 16 Assume A, B are injective MV-algebras. First we realize that the product set

$$A \times B = \{(a, b) \mid a \in A, b \in B\}$$

is an MV-algebra where the MV-operations are defined by

$$(a_1, b_1) \rightarrow (a_2, b_2) = (a_1 \rightarrow a_2, b_1 \rightarrow b_2),\ (a, b)^* = (a^*, b^*).$$

Let C, D be MV-algebras, C is a subalgebra of D and $f : C \searrow A \times B$ is an MV-homomorphism. Clearly, f can be split into two MV-homomorphisms $f_1 : C \searrow A$, $f_2 : C \searrow B$ such that $f(c) = (a, b)$ iff $f_1(c) = a$ and $f_2(c) = b$. Since A, B are injective there are two MV-homomorphisms $h_1 : D \searrow A$, $h_2 : D \searrow B$ such that $h_1(c) = f_1(c)$, $h_2(c) = f_2(c)$ for any $c \in C$. Define a mapping $h : D \searrow A \times B$ by $h(d) = (a, b)$ iff $h_1(d) = a$ and $h_2(d) = b$. Then h is an MV-homomorphism and $h(c) = f(c)$ for any $c \in C$. Therefore $A \times B$ is injective.

Exercise 17 Consider first Łukasiewicz structure. Let $a \in [0, 1]$, n a natural number. Then $b = \frac{a}{n} \in [0, 1]$ is the n-divisor of a. Indeed, $nb = n \cdot \frac{a}{n} = a$, $[a^* \oplus (n-1) \cdot b]^* = 1 - [1 - a + (n-1) \cdot \frac{a}{n}] = b$. Then consider Gaines structure. The only possible way to define an operation $\oplus$ would be $a \oplus b = (a^* \odot b^*)^*$. It is easy too see, however, that for $n = 2$, $nb = \mathbf{0}$ iff $b = \mathbf{0}$ and $nb = \mathbf{1}$ iff $b > \mathbf{0}$. Thus, none of the elements $\mathbf{0} < b < \mathbf{1}$ has a 2-divisor. A similar argument shows that Brouwer structure, too, is not divisible.

5.7 Exercises from Section 2.3

Exercise 1 Take $z = \mathbf{0}$. Then (2.142) is a special case of (2.127). (2.143) is equivalent to $(x \Rightarrow \mathbf{0}) \vee (y \Rightarrow \mathbf{0}) \leq (x \wedge y) \Rightarrow \mathbf{0})$, which holds by (1.66). Since L is a distributive lattice we have $(x^+ \vee y) \wedge x = (x^+ \wedge x) \vee (y \wedge x)$ $= y \wedge x \leq y$. This implies (2.144). Since $(x \Rightarrow y) \wedge y^+ \wedge x \leq y \wedge y^+ \leq \mathbf{0}$, we obtain (2.145) by using the Galois correspondence twice. (2.146) and (2.147) are a special cases of (2.128); take $z = \mathbf{0}$. (2.148) we obtain by using (2.146), (2.140) and (2.146); indeed, $x \Rightarrow y^+ = (x \wedge y)^+ = (x \wedge y)^{+++} = (x \Rightarrow y^+)^{++}$. Finally, since $y \leq y^{++}$, we have $x \Rightarrow y \leq x \Rightarrow y^{++}$, which implies $(x \Rightarrow y^{++})^+ \leq (x \Rightarrow y)^+$ and therefore $(x \Rightarrow y)^{++} \leq (x \Rightarrow y^{++})^{++}$ thus, by (2.148), we have $(x \Rightarrow y)^{++} \leq x \Rightarrow y^{++}$. Hence (2.149) holds.

Exercise 2 (i): Since $a \odot x \leq a \wedge x$ holds for any $a, x \in L$, $a \wedge x = \mathbf{0}$ implies $a \odot x = \mathbf{0}$. Thus, $a^+ = \bigvee\{x \mid a \wedge x = \mathbf{0}\} \leq \bigvee\{x \mid a \odot x = \mathbf{0}\} = a^*$. Let $a \in BoL$. Then $a \odot a = a$ and $a^* \wedge a = (a^* \oplus a^*) \odot a = (a \odot a)^* \odot a = a^* \odot a = \mathbf{0}$, therefore $a^* \leq a \Rightarrow \mathbf{0} = a^+$. Hence $a^+ = a^*$ for any $a \in BoL$.
(ii): holds by (2.135).

5.8 Exercises from Section 3.1

Exercise 1 (a): p, where p stands for *Mary is blond.*
(b): (p `imp` q), where p stands for *Mary is right* and q stands for *Bob is wrong.*
(c): (p `or` `non-`q), where p stands for *Bob made a mistake* and q stands for *There is ball today.*
(d): (p `imp` `non-`q), where p stands for *It is raining* and q stands for *Bob can go to ball by walk.*
(e): (p `imp` q), where p stands for *I'll work through the whole night* and q stands for *I'll be tired.*
(f): [(p `and` q) `or` `non-`r], where p stands for *It is raining,* q stands for *The Sun is shining* and r stands for *We can see the rainbow.*
(g): [(p `or` q) `and` `non-`r], where p stands for *It is raining,* q stands for *It is sunny* and r stands for *It is windy.*

Exercise 2 Let p stand for *Wages rise,* q stand for *Prices rise,* r stand for *There is inflation,* s stand for *The Government stops inflation,* t stand for *People suffer,* w stand for *The Government loses its popularity.* Then
(a): [(p `or` q) `imp` r],
(b): [r `imp` (s `or` t)],
(c): (t `imp` w),
(d): (`non-`s `and` `non-`w)
(e): `non-`p.
Logical connectives `or`, `and` can be replaced also by $\overline{\texttt{or}}$, $\overline{\texttt{and}}$, respectively.

Exercise 3 Let v be a valuation. Then
(a): $v(\texttt{non-}\alpha) = v(\alpha \texttt{ imp } \mathbf{0}) = v(\alpha) \rightarrow v(\mathbf{0}) = v(\alpha) \rightarrow \mathbf{0} = v(\alpha)^*$.

(b):

$$\begin{aligned} v(\alpha \texttt{ or } \beta) &= v(\texttt{non-[non-}\alpha \texttt{ and non-}\beta\texttt{]}) \\ &= v(\texttt{non-}\alpha \texttt{ and non-}\beta)^* \\ &= [v(\texttt{non-}\alpha) \odot v(\texttt{non-}\beta)]^* \\ &= [v(\alpha)^* \odot v(\beta)^*]^* \\ &= [v(\alpha) \oplus v(\beta)]. \end{aligned}$$

Exercise 4 (a): Any valuation v which satisfies T has a property

$$v(\texttt{non-}s \texttt{ and } \texttt{non-}w) = v(s)^* \odot v(w)^* = 1,$$

thus $v(s) \oplus v(w) = 0$ and therefore $v(s) = v(w) = 0$. Since $v(t \texttt{ imp } w) = v(t) \rightarrow v(w) = v(t) \rightarrow 0 \geq 0.8$, we conclude $0 \leq v(t) \leq 0.2$ for any v satisfying T. Study first the case $v_1(t) = 0.2$. Then

$$\begin{aligned} v_1(r \texttt{ imp } (s \texttt{ or } t)) &= v_1(r) \rightarrow [v_1(s) \oplus v_1(t)] \\ &= v_1(r) \rightarrow (0 \oplus 0.2) \\ &= 1 - v_1(r) + 0.2 \\ &\geq 0.9. \end{aligned}$$

Thus, $v_1(r) \leq 0.3$. Now $v(\texttt{non-}p)$ obtains the smallest value when $v(p)$ obtains the largest value, and, by assumption, $v(p) \oplus v(q) \leq v(r)$. Take therefore $v_1(q) = 0$, $v_1(r) = 0.3$, then $v_1(p) \leq 0.3$, in particular, take $v_1(p) = 0.3$. Then $v_1(\texttt{non-}p) = 0.7$. Clearly, v_1 satisfies T. Next, study the case $v_2(t) = 0$. Since $0.9 \leq v_2(r) \rightarrow [v_2(s) \oplus v_2(t)] = v_2(r)^*$, we have $v_2(r) \leq 0.1$. Again, $v_2(p) \oplus v_2(q) \leq v_2(r)$, so take $v_2(r) = 0.1$, $v_2(q) = 0$, then $v_2(p) \leq 0.1$, thus $0.9 \leq v_2(\texttt{non-}p)$. We conclude that $C^{\mathbf{sem}}(T)(\texttt{non-}p) = v_1(\texttt{non-}p) = 0.7$.
(b): Replace or by $\overline{\texttt{or}}$ and and by $\overline{\texttt{and}}$. Any valuation v which satisfies T necessarily fulfils $v(s) = v(w) = 0$. Again we reason that $v(\texttt{non-}p)$ obtains the smallest value when $v(t) = 0.2$, $v(r) = 0.3$ and $v(q) = 0$. Then $v(p) \vee v(q) \leq v(r)$ and therefore $v(p) = 0.3$. This implies $v(\texttt{non-}p) = C^{\mathbf{sem}}(T)(\texttt{non-}p) = 0.7$.

Exercise 5 Let $v(\alpha), v(\beta) \in \{0, 1\}$. Then we have $v(\alpha \texttt{ or } \beta) = v(\alpha) \oplus v(\beta) = 0$ iff $v(\alpha) = v(\beta) = 0$ iff $v(\alpha) \vee v(\beta) = v(\alpha\ \overline{\texttt{or}}\ \beta) = 0$ and $v(\alpha \texttt{ or } \beta) = 1 = v(\alpha\ \overline{\texttt{or}}\ \beta)$ elsewhere. Similarly, $v(\alpha \texttt{ and } \beta) = v(\alpha) \odot v(\beta) = 1$ iff $v(\alpha) = v(\beta) = 1$ iff $v(\alpha) \wedge v(\beta) = v(\alpha\ \overline{\texttt{and}}\ \beta) = 1$ and $v(\alpha \texttt{ and } \beta) = 0 = v(\alpha\ \overline{\texttt{and}}\ \beta)$ elsewhere. This proves (b). To establish (a) it enough to realize that $v(\alpha \texttt{ imp } \beta) = v(\alpha) \rightarrow v(\beta) = 0$ iff $v(\alpha) = 1$, $v(\beta) = 0$ and otherwise $v(\alpha \texttt{ imp } \beta) = 1$. In particular, $v(\texttt{ non-}\alpha) = v(\alpha)^* = 1$ iff $v(\alpha) = 0$ and $v(\alpha)^* = 0$ iff $v(\alpha) = 1$.

Exercise 6

	$v(\alpha$ and $\beta)$	$v(\beta)$ 0.0	0.20	0.4	0.6	0.8	1.0
	0.0	0.0	0.0	0.0	0.0	0.0	0.0
	0.2	0.0	0.0	0.0	0.0	0.0	0.2
$v(\alpha)$	0.4	0.0	0.0	0.0	0.0	0.2	0.4
	0.6	0.0	0.0	0.0	0.2	0.4	0.6
	0.8	0.0	0.0	0.2	0.4	0.6	0.8
	1.0	0.0	0.20	0.4	0.6	0.8	1.0

	$v(\alpha$ or $\beta)$	$v(\beta)$ 0.0	0.2	0.4	0.6	0.8	1.0
	0.0	0.0	0.2	0.4	0.6	0.8	1.0
	0.2	0.2	0.4	0.6	0.8	1.0	1.0
$v(\alpha)$	0.4	0.4	0.6	0.8	1.0	1.0	1.0
	0.6	0.6	0.8	1.0	1.0	1.0	1.0
	0.8	0.8	1.0	1.0	1.0	1.0	1.0
	1.0	1.0	1.0	1.0	1.0	1.0	1.0

	$v(\alpha$ imp $\beta)$	$v(\beta)$ 0.0	0.2	0.4	0.6	0.8	1.0
	0.0	1.0	1.0	1.0	1.0	1.0	1.0
	0.2	0.8	1.0	1.0	1.0	1.0	1.0
$v(\alpha)$	0.4	0.6	0.8	1.0	1.0	1.0	1.0
	0.6	0.4	0.6	0.8	1.0	1.0	1.0
	0.8	0.2	0.4	0.6	0.8	1.0	1.0
	1.0	0.0	0.2	0.4	0.6	0.8	1.0

Exercise 7 In Classical Propositional Logic, a formula α is a tautology iff $v(\alpha) = 1$. Only in this case $\mathcal{C}^{\mathbf{sem}}(\alpha) = 1$.

Exercise 8 All equations (3.9)-(3.19) are direct consequences of the definition of truth value function and properties valid in residuated lattice. For example, by (1.45),

$$v((\alpha \texttt{imp} \beta) \texttt{ imp } [(\beta \texttt{imp} \gamma) \texttt{ imp } (\alpha \texttt{imp} \gamma)]) =$$
$$[v(\alpha) \to v(\beta)] \to \{[v(\beta) \to v(\gamma)] \to [v(\alpha) \to v(\gamma)]\} = 1.$$

Therefore, (3.10) holds.

Exercise 9 For example

$$p \texttt{ imp } \{\{q \texttt{ imp } [(r \texttt{ imp } \mathbf{0}) \texttt{ imp } (s \texttt{ imp } \mathbf{0})]\} \texttt{ imp } \mathbf{0}\}$$

is such a formula. The first connective is here the main connective.

Exercise 10 Since

$$\begin{aligned} v(\alpha \texttt{ xor } \beta) &= [v(\alpha) \oplus v(\beta)] \wedge [v(\alpha) \to v(\beta)^*] \wedge [v(\beta) \to v(\alpha)^*] \\ &= [v(\alpha) \oplus v(\beta)] \wedge [v(\alpha)^* \oplus v(\beta)^*] \wedge [v(\beta)^* \oplus v(\alpha)^*] \\ &= [v(\alpha) \oplus v(\beta)] \wedge [v(\alpha)^* \oplus v(\beta)^*], \\ &= [v(\alpha) \oplus v(\beta)] \wedge [v(\alpha) \odot v(\beta)]^* \end{aligned}$$

we have (a), and by the last line, (c) is now obvious. Finally, (b):

		$v(\beta)$					
	$v(\alpha \texttt{ xor } \beta)$	0.0	0.2	0.4	0.6	0.8	1.0
	0.0	0.0	0.2	0.4	0.6	0.8	1.0
	0.2	0.2	0.4	0.6	0.8	1.0	0.8
$v(\alpha)$	0.4	0.4	0.6	0.8	1.0	0.8	0.6
	0.6	0.6	0.8	1.0	0.8	0.6	0.4
	0.8	0.8	1.0	0.8	0.6	0.4	0.2
	1.0	1.0	0.8	0.6	0.4	0.2	0.0

Exercise 11 Let p stand for *Bob reads for the examination*, q stand for *Bob goes for beer*, r stand for *Bob wants to meet girls*, s stand for *Bob passes the examination.* Then
(a) stands for $(p \texttt{ xor } q)$,
(b) stands for $(p \texttt{ imp } s)$,
(c) stands for $[r \texttt{ imp } (\texttt{non} - p)]$,
(d) stands for $(r \texttt{ and } q)$,
(e) stands for s.

Exercise 12 Any valuation v which satisfies T fulfils

$$v(r \texttt{ and } q) = v(r) \odot v(q) = 1,$$

thus $v(r) = v(q) = 1$ and $0.8 \leq v(p \texttt{ xor } q) = [v(p) \oplus v(q)] \wedge [v(p)^* \oplus v(q)^*] = [v(p) \oplus 1] \wedge [v(p)^* \oplus 0] = v(p)^*$. Therefore $0 \leq v(p) \leq 0.2$. Moreover, $0.7 \leq v(r \texttt{ imp } (\texttt{non} - p)) = v(r) \to v(p)^* = 1 \to v(p)^* = v(p)^*$. Thus, $0 \leq v(p) \leq 0.3$. Since $0.8 \leq v(p \texttt{ imp } s) = v(p) \to v(s)$ iff $0.8 \odot v(p) \leq v(s)$ we realize that $v(s)$ obtains its smallest value when $v(p) = 0$. In that case $v(s) = 0$. We conclude $C^{sem}(T)(s) = 0$. Poor Bob! Let us hope he at least has a nice ball with girls in pub!

Exercise 13 A valuation v such that $v(r) = v(q) = v(s) = 1$ and $v(p) = 0$ satisfies T and $v(\texttt{non}-s) = 0$. Therefore also $\texttt{non}-s$ is valid at the degree 0. This holds also generally. From the fact that α is valid at the degree a we cannot derive $\texttt{non}-\alpha$ is valid at the degree a^*!

5.9 Exercises from Section 3.2

Exercise 1 Define $R : [0,1] \searrow [0,1]$ by $R(1) = 1$ and, for each $a \in [0,1)$, $R(a) = 0$. Then clearly R is isotone. On the other hand, $\bigvee_{a\in[0,1)} R(a) = 0 \neq R(\bigvee\{a \mid a \in [0,1)\}) = R(1) = 1$.

Exercise 2 No! Since L is a finite totally-ordered lattice, it is enough to study the case $a \leq b$. Since R is isotone, $R(a) \leq R(b)$. Thus, $R(a \vee b) = R(b) = R(b) \vee R(a)$. Therefore R is semi-continuous.

Exercise 3 (a): Let w_1 be a metaproof for α and w_2 be a metaproof for $(\alpha \texttt{ imp } \beta)$ such that $Val_{\langle \mathtt{A},\mathtt{R},T\rangle}(w_1) = c$ and $Val_{\langle \mathtt{A},\mathtt{R},T\rangle}(w_2) = b$. Combining these two metaproofs we obtain the following metaproof for β:

α	,	c	,	assumption
$\alpha \texttt{ imp } \beta$	,	b	,	assumption
β	,	$c \odot b$	,	R_{RC}

Thus, $a \odot b \leq C^{\mathtt{syn}\langle \mathtt{A},\mathtt{R}\rangle}(T)(\beta)$. By (1.28), we thus have

$$C^{\mathtt{syn}\langle \mathtt{A},\mathtt{R}\rangle}(T)(\alpha) \odot C^{\mathtt{syn}\langle \mathtt{A},\mathtt{R}\rangle}(T)(\alpha \texttt{ imp } \beta) \leq C^{\mathtt{syn}\langle \mathtt{A},\mathtt{R}\rangle}(T)(\beta).$$

(b): Similarly, we have the following metaproof for $(\mathbf{a} \texttt{ imp } \alpha)$:

α	,	c	,	assumption
$\mathbf{a}$	,	a	,	Ax.7
$\mathbf{a} \texttt{ imp } \alpha$	,	$a \to c$	,	$\mathrm{R}_{\mathbf{a}-\mathrm{LR}}$

Therefore $a \to c \leq C^{\mathtt{syn}\langle \mathtt{A},\mathtt{R}\rangle}(T)(\mathbf{a} \texttt{ imp } \alpha)$. By (1.67), we conclude

$$\mathbf{a} \to C^{\mathtt{syn}\langle \mathtt{A},\mathtt{R}\rangle}(T)(\alpha) \leq C^{\mathtt{syn}\langle \mathtt{A},\mathtt{R}\rangle}(T)(\mathbf{a} \texttt{ imp } \alpha).$$

(c): Consider the following metaproof for $(\alpha \textbf{ and } \beta)$:

α	,	c	,	assumption
β	,	b	,	assumption
$\alpha \textbf{ and } \beta$	,	$c \odot b$	,	$\mathrm{R}_{\mathrm{RBC}}$

Hence $c \odot b \leq C^{\mathtt{syn}\langle \mathtt{A},\mathtt{R}\rangle}(T)(\alpha \textbf{ and } \beta)$. Similarly to (a), we thus have

$$C^{\mathtt{syn}\langle \mathtt{A},\mathtt{R}\rangle}(T)(\alpha) \odot C^{\mathtt{syn}\langle \mathtt{A},\mathtt{R}\rangle}(T)(\beta) \leq C^{\mathtt{syn}\langle \mathtt{A},\mathtt{R}\rangle}(T)(\alpha \textbf{ and } \beta).$$

Exercise 4 The product operation $\odot$ of any residuated lattice is isotone and fulfils $\mathbf{0} \odot a = \mathbf{0}$, $\mathbf{1} \odot \mathbf{1} = \mathbf{1}$. Thus (MP4), (MP3) and (MP2), respectively, hold. Given a valuation v and formulas α, β, we have $v(\alpha) \odot v(\alpha \texttt{ imp } \beta) = v(\alpha) \odot [v(\alpha) \to v(\beta)] \leq v(\beta)$. Hence, (MP1) holds.

Exercise 5 The complement operation * of any MV-algebra defined on the unit is a continuous function $^* : [0,1] \searrow [0,1]$ as the operation $\to$ is continuous and $a^* = a \to \mathbf{0}$. Since, for each $a, b \in [0,1]$, $a \leq b$ implies $b^* \leq a^*$, we see that * is decreasing. It is also strictly decreasing as if $a < b$

would imply $b^* = a^*$, then $a = a^{**} = b^{**} = b$, a contradiction. (N1) - (N3) read $a^{**} = a$, $1^* = 0$, $0^* = 1$, valid in MV-algebras.

Exercise 6 For the residual operation $\to$ of any residuated lattice defined on the unit interval, for each $a, b, c \in [0,1]$, if $a \leq b$, then $b \to c \leq a \to c$ and $c \to a \leq c \to b$, thus (I1) and (I2). Moreover, $\mathbf{0} \to a = \mathbf{1}$, $\mathbf{1} \to a = a$, and $a \to (b \to c) = b \to (a \to c)$. Thus (I3) - (I.5), respectively, are valid.

Exercise 7 For each valuations v and all formulas α, β holds

$$v(\texttt{non-}\beta) \odot v(\alpha \texttt{ imp } \beta) = v(\beta)^* \odot [v(\alpha) \to v(\beta)] \leq v(\alpha)^* = v(\texttt{non-}\alpha),$$

thus (MT1) holds. (MT2), (MT3) and (MP4) are obvious.

Exercise 8 Take $n =^*$. Then, for each $a, b \in [0,1]$,

$$b^* \to a^* \leq a \to b = a^{**} \to b^{**} \leq b^* \to a^*$$

Thus, $a \to b = b^* \to a^*$.

Exercise 9 The product operation $\odot$ of any residuated lattice defined on the unit interval is, by Exercise 1.3.16, a t-norm. In any complete MV-algebra, the operation $\odot$ is, by Exercise 2.2.6, a continuous function. By (1.29), for each $a, b \in [0,1]$,

$$a \to b = \bigvee\{c \mid a \odot c \leq b\}.$$

Therefore the residual operation $\to$ of any complete MV-algebra defined on the unit interval is an R-implication.

Exercise 10 (a): In any complete MV-algebra holds (1.65). Thus, Rule of Bold Disjunction is semi-continuous as

$$r^{\texttt{sem}}(a, \bigvee_{i\in\Gamma} b_i) = a \oplus (\bigvee_{i\in\Gamma} b_i) = \bigvee_{i\in\Gamma}(a \oplus b_i) = \bigvee_{i\in\Gamma} r^{\texttt{sem}}(a, b_i).$$

Given any valuation v, we reason soundness by observing that

$$\begin{aligned} r^{\texttt{sem}}(v(\alpha), v(\beta)) &= v(\alpha) \oplus v(\beta) \\ &= v(\alpha \texttt{ or } \beta) \\ &\leq v(r^{\texttt{syn}}(\alpha, \beta)). \end{aligned}$$

(b): In any residuated lattice holds (1.17). Thus, Rule of Conjunction is semi-continuous as

$$r^{\texttt{sem}}(a, \bigvee_{i\in\Gamma} b_i) = a \wedge (\bigvee_{i\in\Gamma} b_i) = \bigvee_{i\in\Gamma}(a \wedge b_i) = \bigvee_{i\in\Gamma} r^{\texttt{sem}}(a, b_i).$$

Given any valuation v, we have soundness as

$$\begin{aligned} r^{\texttt{sem}}(v(\alpha), v(\beta)) &= v(\alpha) \wedge v(\beta) \\ &= v(\alpha \ \overline{\texttt{and}}\ \beta) \\ &\leq v(r^{\texttt{syn}}(\alpha, \beta)). \end{aligned}$$

Exercise 11 Soundness: equally to the case of Generalized Modus

Ponens. Semi-continuity: Let v be a valuation. Recall that

$$\begin{aligned} v(\alpha \ \mathtt{xor} \ \beta) &= [v(\alpha) \oplus v(\beta)] \wedge [v(\alpha)^* \oplus v(\beta)^*] \\ &\leq [v(\alpha) \oplus v(\beta)] \\ &= [v(\alpha)^* \to v(\beta)] \\ &= [v(\mathtt{non} - \alpha \ \mathtt{imp} \ \beta)]. \end{aligned}$$

Therefore $v(\alpha \ \mathtt{xor} \ \beta) \odot v(\mathtt{non}\text{-}\alpha) \leq v(\beta)$. Thus,

$$\begin{aligned} r^{\mathtt{sem}}(v(\alpha \ \mathtt{xor} \ \beta), v(\mathtt{non}\text{-}\alpha)) &= v(\alpha \ \mathtt{xor} \ \beta) \odot v(\mathtt{non}\text{-}\alpha) \\ &\leq v(\beta) \\ &= v(r^{\mathtt{sym}}(\alpha \ \mathtt{xor} \ \beta), \mathtt{non}\text{-}\alpha). \end{aligned}$$

Exercise 12 Besides non-logical axioms $T(p \ \mathtt{xor} \ q) = 1, T(p \ \mathtt{imp} \ s) = 1$ and $T(\mathtt{non} - q) = 0.9$ we need just the fuzzy rules R_{GMP} and R_{XOR}. By choosing a valuation v such that $v(p) = v(s) = 0.9$, $v(q) = 0.1$ we reason that this fuzzy theory is satisfiable, thus consistent. Moreover, we have the following metaproof for s:

$p \ \mathtt{xor} \ q$	1	non-logical axiom
$\mathtt{non} - q$	0.9	non-logical axiom
p	0.9	R_{XOR}
$p \ \mathtt{imp} \ s$	1	non-logical axiom
s	0.9	R_{GMP}

Thus, $C^{\mathtt{syn}}(T)(s) = 0.9$. Freely speaking, *Bob passes the examination but not with best grade.*

5.10 Exercises from Section 3.3

Exercise 1 Assume $\alpha \preceq \beta$, $\beta \preceq \gamma$. Then $T \vdash_1 (\alpha \ \mathtt{imp} \ \beta)$, $T \vdash_1 (\beta \ \mathtt{imp} \ \gamma)$. Let w_1 and w_2 be metaproofs for $(\alpha \ \mathtt{imp} \ \beta)$ and $(\beta \ \mathtt{imp} \ \gamma)$, respectively, such that $Val_T(w_1) = a$ and $Val_T(w_2) = b$. By combining these two metaproofs we obtain the following metaproof w_3 for $(\alpha \ \mathtt{imp} \ \gamma)$:

$(\alpha \ \mathtt{imp} \ \beta)$	a	assumption
$(\alpha \ \mathtt{imp} \ \beta) \ \mathtt{imp} \ [(\beta \ \mathtt{imp} \ \gamma) \ \mathtt{imp} \ (\alpha \ \mathtt{imp} \ \gamma)]$	1	Ax.2
$(\beta \ \mathtt{imp} \ \gamma) \ \mathtt{imp} \ (\alpha \ \mathtt{imp} \ \gamma)$	a	R_{GMP}
$(\beta \ \mathtt{imp} \ \gamma)$	b	assumption
$(\alpha \ \mathtt{imp} \ \gamma)$	$a \odot b$	R_{GMP}

Since $a \odot b = Val_T(w_3) \leq C^{\mathtt{syn}}(T)(\alpha \ \mathtt{imp} \ \gamma)$, we have $T \vdash_1 (\alpha \ \mathtt{imp} \ \gamma)$. Therefore $\alpha \preceq \gamma$, whence $\preceq$ is transitive.

Exercise 2 Assume $\alpha_1 \sim \beta_1$ and $\alpha_2 \sim \beta_2$. Then $T \vdash_1 (\alpha_1 \ \mathtt{imp} \ \beta_1)$ and $T \vdash_1 (\beta_2 \ \mathtt{imp} \ \alpha_2)$. Let w_1 and w_2 be metaproofs for $(\alpha_1 \ \mathtt{imp} \ \beta_1)$ and $(\beta_2 \ \mathtt{imp} \ \alpha_2)$, respectively, and $Val_T(w_1) = a$, $Val_T(w_2) = b$. Then we have the following metaproof for a formula

$$\gamma = [(\beta_1 \texttt{ imp } \beta_2) \texttt{ imp } (\alpha_1 \texttt{ imp } \alpha_2)].$$

$(\alpha_1 \texttt{ imp } \beta_1)$	a	assumption
$(\alpha_1 \texttt{ imp } \beta_1) \texttt{ imp } [(\beta_2 \texttt{ imp } \alpha_2) \texttt{ imp } \gamma]$	1	Ax.3
$[(\beta_2 \texttt{ imp } \alpha_2) \texttt{ imp } \gamma]$	a	R_{GMP}
$(\beta_2 \texttt{ imp } \alpha_2)$	b	assumption
γ	$a \odot b$	R_{GMP}

We conclude $T \vdash_1 [(\beta_1 \texttt{ imp } \beta_2) \texttt{ imp } (\alpha_1 \texttt{ imp } \alpha_2)]$. Similarly, $T \vdash_1 [(\alpha_1 \texttt{ imp } \alpha_2) \texttt{ imp } (\beta_1 \texttt{ imp } \beta_2)]$. We conclude

$$(\alpha_1 \texttt{ imp } \alpha_2) \sim (\beta_1 \texttt{ imp } \beta_2).$$

Thus, $\sim$ is a congruence with respect to the logical connective imp. Now assume $\alpha \sim \beta$. Since $\mathbf{0} \sim \mathbf{0}$, we have $(\alpha \texttt{ imp } \mathbf{0}) \sim (\beta \texttt{ imp } \mathbf{0})$, whence $(\texttt{non-}\alpha) \sim (\texttt{non-}\beta)$, i.e. $\sim$ is a congruence with respect to the logical connective non.

Exercise 3 Let $|\alpha| \in (\mathcal{F}/\sim)$. By Ax.5 and Ax.4, $T \vdash_1 (\mathbf{0} \texttt{ imp } \alpha)$ and $T \vdash_1 (\alpha \texttt{ imp } \mathbf{1})$ thus, $|\mathbf{0}| \leq |\alpha| \leq |\mathbf{1}|$, accordingly, $|\mathbf{0}|$ and $|\mathbf{1}|$ are the least element and the greatest element in $\mathcal{F}/\sim$.

Exercise 4 $T \vdash_a \alpha$ implies $T \vdash_1 (\mathbf{a} \texttt{ imp } \alpha)$ thus, $|\mathbf{a} \texttt{ imp } \alpha| = |\mathbf{1}|$. Since h is MV-homomorphism it maps unit to unit, accordingly, $\mathbf{1} = h(|\mathbf{a} \texttt{ imp } \alpha|) = h(|\mathbf{a}|) \to h(|\alpha|)$, whence $h(|\mathbf{a}|) \leq h(|\alpha|)$.

Exercise 5 The claim follows easily by definition of these logical connectives and the corresponding MV-algebra operations.

Exercise 6 Let α, β, γ, δ be formulas and v a valuation. Then it is easy to see that

$$[v(\alpha) \oplus v(\beta)] \odot [v(\alpha) \to v(\gamma)] \odot [v(\beta) \to v(\delta)] \leq v(\gamma) \oplus v(\delta).$$

Thus,

$$\begin{aligned} r^{sem}(v(\alpha \texttt{or} \beta), v(\alpha \texttt{imp} \gamma), v(\beta \texttt{or} \delta)) &\leq v(\gamma) \oplus v(\delta) \\ &= v(\gamma \texttt{ or } \delta) \\ &= v(r^{sem}(\alpha \texttt{or} \beta), (\alpha \texttt{imp} \gamma), (\beta \texttt{or} \delta)). \end{aligned}$$

Thus, Generalized Disjunctive Syllogism is sound.

Exercise 7 We let p stand for *A is regularly in a hurry*, q stand for *A is always nervous*, r stand for *A has bad eating habits*, s stand for *A has peritonitis*, t stand for *A contracts gastric ulcer*, w stand for *A is old*, z stand for *The illness is fatal.*

Then the fuzzy theory T can be expressed by setting $T(p \texttt{ imp } q) = 0.8$,[1] $T(r \texttt{ imp } s) = 0.8$, $T((q \texttt{ or } s) \texttt{ imp t}) = 1$, $T(w \texttt{ imp } (t \texttt{ imp z})) = 1$, $T(w) = 0.6$ [sic!], $T(p) = 1$, $T(r) = 1$.

First we are looking for a metaproof for the formula t. We have

p	1	non-logical axiom
r	1	non-logical axiom
$(p \texttt{ or } r)$	1	R_{RBD}
$(p \texttt{ imp } q)$	0.8	non-logical axiom
$(r \texttt{ imp } s)$	0.8	non-logical axiom
$(q \texttt{ or } s)$	0.6	R_{GDS}
$[(q \texttt{ or } s) \texttt{ imp t}]$	1	non-logical axiom
t	0.6	R_{GMP}

Thus, the degree of deduction of the formula t is at least 0.6. We have, however, also the following metaproof for t and z:

p	1	non-logical axiom
$(p \texttt{ imp } q)$	0.8	non-logical axiom
q	0.8	R_{GMP}
r	1	non-logical axiom
$(r \texttt{ imp } s)$	0.8	non-logical axiom
s	0.8	R_{GMP}
$(q \texttt{ or } s)$	1	R_{RBD}
$[(q \texttt{ or } s) \texttt{ imp t}]$	1	non-logical axiom
t	1	R_{GMP}
w	0.6	non-logical axiom
$[w \texttt{ imp } (t \texttt{ imp z})]$	1	non-logical axiom
$(t \texttt{ imp z})$	0.6	R_{GMP}
z	0.6	R_{GMP}

We conclude $\mathcal{C}^{\mathtt{syn}}(T)(t) = 1$ and $0.6 \leq \mathcal{C}^{\mathtt{syn}}(T)(z)$. In semantic side we are looking for valuations satisfying T. Take v such that $v(p) = v(r) = v(t) = 1$, $v(q) = v(s) = 0.8$ $v(w) = v(z) = 0.6$. This valuation satisfies T. Moreover, $\mathcal{C}^{\mathtt{syn}}(T)(z) = \mathcal{C}^{\mathtt{sem}}(T)(z) = 0.6$. Freely speaking, *A middle-aged person who is always in a hurry and has bad eating habits will contract gastric ulcer but the illness is less fatal.*

Exercise 8 Informally we would reason *Since Mary wears jeans she is relaxed. Thus, Bob likes her, whence Mary makes an impression on Bob. Therefore the story has a happy end.* Formally we draw the same conclusion in the following way.

[1] Notice that, if we would have *Nervousness* always *ensues from precipitation* then we would write $T(p \texttt{ imp } q) = 1$

We let a stand for *Mary is wearing basic black*, b stand for *Mary is wearing high-heeled shoes*, c stand for *Mary is wearing jacket*, d stand for *Mary is wearing low-heeled shoes*, e stand for *Mary is wearing peaked cap*, f stand for *Mary is wearing jeans*, g stand for *Mary is wearing tennis shoes*, A stand for *Mary looks tempting*, B stand for *Mary looks academic*, C stand for *Mary is relaxed*, p stand for *Bob is sober*, q stand for *Bob likes Mary*, r stand for *Happy End*.

Then we define a fuzzy theory T by setting
$T([a \text{ and } b] \text{ imp } A) = 1$,
$T([c \text{ and } d] \text{ imp } B) = 1$,
$T([e \text{ or } f \text{ or } g] \text{ imp } C) = 1$,
$T([p \text{ and } (\text{non} - A)] \text{ imp } [\text{non} - q]) = 1$,
$T([A \text{ and } (\text{non} - p)] \text{ imp } q) = 1$,
$T([B \text{ and } (\text{non} - p)] \text{ imp } [\text{non} - q]) = 1$,
$T([B \text{ and } p] \text{ imp } q) = 1$,
$T([p \text{ or } (\text{non} - p) \text{ and } C] \text{ imp } q) = 1$,
$T(q \text{ imp } r) = 1$,
$T(b) = T(c) = T(f) = 1$,
$T(p) = T(\text{non} - p) = 0.5$ and $T(\alpha) = 0$ elsewhere.

This theory is satisfiable, indeed, a valuation v such that $T(a) = T(d) = T(e) = T(g) = T(A) = T(B) = 0$, $T(b) = T(c) = T(f) = T(q) = T(r) = T(C) = 1$, $T(p) = 0.5$ satisfies T. The following is a metaproof for r:

e	, 0 ,	non-logical axiom
f	, 1 ,	non-logical axiom
$(e \text{ or } f)$	, 1 ,	R_{RBD}
g	, 0 ,	non-logical axiom
$(e \text{ or } f \text{ or } g)$	, 1 ,	R_{RBD}
$(e \text{ or } f \text{ or } g) \text{ imp } C$	, 1 ,	non-logical axiom
C	, 1 ,	R_{GMP}
p	, 0.5 ,	non-logical axiom
$\text{non} - p$	, 0.5 ,	non-logical axiom
$(p \text{ or } [\text{non} - p])$	, 1 ,	R_{RBD}
$(p \text{ or } [\text{non} - p]) \text{ and } C$	, 1 ,	R_{RBC}
$[(p \text{ or non} - p) \text{ and } C] \text{ imp } q$	, 1 ,	non-logical axiom
q	, 1 ,	R_{GMP}
$(q \text{ imp } r)$	, 1 ,	non-logical axiom
r	, 1 ,	R_{GMP}

Consequently, $T \vdash_1$ *Happy End*.

5.11 Exercises from Section 4.1

Exercise 1 Since the operation μ on L is isotone the μ-composition of fuzzy relations is isotone. Moreover, since μ on L is associative and preserves joins in L we have, for each $(u, z) \in U \times Z$,

$$\begin{aligned} R\mu(S\mu T)\langle u,z\rangle &= \bigvee_{v\in V} \mu(R\langle u,v\rangle, \bigvee_{w\in W} \mu(S\langle v,w\rangle, T\langle w,z\rangle)) \\ &= \bigvee_{v\in V, w\in W} \mu(R\langle u,v\rangle, \mu(S\langle v,w\rangle, T\langle w,z\rangle)) \\ &= \bigvee_{v\in V, w\in W} \mu(\mu(R\langle u,v\rangle, S\langle v,w\rangle), T\langle w,z\rangle) \\ &= \bigvee_{w\in W} \mu(\bigvee_{v\in V} \mu(R\langle u,v\rangle, S\langle v,w\rangle), T\langle w,z\rangle) \\ &= (R\mu S)\mu T\langle u,z\rangle, \end{aligned}$$

therefore $R\mu(S\mu T) = (R\mu S)\mu T$.

Exercise 2 In the case of μ commutative we have $f = g$ and for each $(w, u) \in W \times U$,

$$\begin{aligned} S^{-1}\mu R^{-1}\langle w,u\rangle &= \bigvee_{v\in V} \mu(S^{-1}\langle w,v\rangle, R^{-1}\langle v,u\rangle) \\ &= \bigvee_{v\in V} \mu(R^{-1}\langle v,u\rangle, S^{-1}\langle w,v\rangle) \\ &= \bigvee_{v\in V} \mu(R\langle u,v\rangle, S\langle v,w\rangle) \\ &= R\mu S\langle u,w\rangle \\ &= (R\mu S)^{-1}\langle w,u\rangle. \end{aligned}$$

Thus, $S^{-1}\mu R^{-1} = (R\mu S)^{-1}$. Moreover, for each $(v, w) \in V \times W$,

$$\begin{aligned} H_1(R^{-1},T^{-1})\langle v,w\rangle &= \bigwedge_{u\in u} f(R^{-1}\langle v,u\rangle, T^{-1}\langle w,u\rangle) \\ &= \bigwedge_{u\in u} g(R^{-1}\langle v,u\rangle, T^{-1}\langle w,u\rangle) \\ &= \bigwedge_{u\in u} g(R\langle u,v\rangle, T\langle u,w\rangle) \\ &= H_2(R,T)\langle v,w\rangle, \end{aligned}$$

implying $H_1(R^{-1},T^{-1}) = H_2(R,T)$.

Exercise 3 If $C = \bigcap_{i=1}^{n} H_1(S_i, Ti)$ is a solution of the system of fuzzy relation equations $(*)$ $X\mu S_i = T_i$, $i = 1, \cdots, n$, then $(*)$ has a solution. Conversely, if $(*)$ has a solution, then each fuzzy relation equation $X\mu S_i = T_i$ has a solution and $H_1(S_i, Ti)$ is the largest solution in each case. Then C is a solution of $(*)$ and is obviously the largest one.

Exercise 4 If a solution exists, then

$H_1(S,T)\langle u,v\rangle =$

	v_1	v_2	v_3
u_1	0.5	0.6	0.5
u_2	0.3	0.4	0.3
u_3	0.9	1.0	1.0

is a solution, however, $H_1(S,T)\mu S\langle u_1, w_1\rangle = 0 \neq T\langle u_1, w_1\rangle$, therefore no solution exists.

Exercise 5

$R\mu S\langle u,w\rangle =$

	w_1	w_2	w_3
u_1	0.18	0.9	0.45
u_2	0.12	0.6	0.3
u_3	0.06	0.3	0.15

where, for $i,j=1,2,3$, $R\mu S\langle u_i,w_j\rangle = \max_{v\in V}\{R\langle u_i,v\rangle \cdot S\langle v,w_j\rangle\}$.

Exercise 6

$H_1(S,T)\langle u,v\rangle =$

	v_1	v_2	v_3
u_1	0.9	1	0.9
u_2	0.6	$\frac{2}{3}$	0.6
u_3	0.3	$\frac{1}{3}$	0.3

where, for $i,j=1,2,3$, $H_1(S,T)\langle u_i,v_j\rangle = \min_{w\in W}\{\frac{T\langle u_i,w\rangle}{S\langle v_j,w\rangle} \wedge 1\}$.

Exercise 7

$$T \underset{\sim}{\subset} \texttt{SYMPTOMS} \times \texttt{MEDICAL CARE} =$$

	rest	preg.test	painkiller	outdoor l.
headache	0.9	0	0.9	0.6
nausea	0.5	1	0.2	0.5
constipation	1	0	0.1	1
diarrhea	0.4	0	0.4	0.1
fever	1	0	0.8	0
erythema	1	0	0.4	1

The relation tells us how intensive treatment one has to order to each symptom.

Exercise 8 The maximal solution is

$$H_1(S,T) \underset{\sim}{\subset} \texttt{SYMPTOMS} \times \texttt{ILLNESSES} =$$

	hangover	pregnancy	flu	overexertion
headache	0.9	0	0.9	0.6
nausea	0.2	1	0.4	0.5
constipation	0.1	0	0.3	1
diarrhea	0.4	0	0.4	0.1
fever	0.3	0	1	0
erythema	0.4	0	0.6	1

The relation tells the maximal intensity the symptoms can grow without changing medical care.

5.12 Exercises from Section 4.2

Exercise 1 Let S be a fuzzy equivalence relation based on the Łukasiewicz product. Then, for each $x, y, z \in [0,1]$, $1 - S\langle x, y\rangle = 1 - S\langle y, x\rangle$ and $1 - S\langle x, z\rangle \leq \{1 - S\langle x, y\rangle + 1 - S\langle y, z\rangle\} \wedge 1$ iff $\{S\langle x, y\rangle + S\langle y, z\rangle - 1\} \vee 0$ which holds true. Moreover, $1 - S\langle x, x\rangle = 0$. Therefore $1 - S$ is a pseudo-metric bounded by 1. The converse holds by a symmetric argument.

Exercise 2 Since, for each $x, y, z \in [0,1]$, $S\langle x, y\rangle \wedge S\langle y, z\rangle \leq S\langle x, z\rangle$ iff $1 - S\langle x, z\rangle \leq \max\{1 - S\langle x, y\rangle, 1 - S\langle y, z\rangle$ it is clear that, given a fuzzy similarity relation S based on the Brouwer product, $1 - S$ is an ultra-metric bounded by 1, and vice versa.

5.13 Exercises from Section 4.3

Exercise 1 Case (i) $\Box = \rightarrow$ and $\sqcap = \odot$. Taking the normality of μ_A and (1.70) into account, we obtain

$$\begin{aligned}(\mu_A \circ_\sqcap R_*)(y) &= \bigvee_{x \in X} \{\mu_A(x) \odot [\mu_A(x) \rightarrow \mu_B(y)]\} \\ &= \bigvee_{x \in X} \{\mu_A(x) \wedge \mu_B(y)\} \\ &= \mu_B(y).\end{aligned}$$

Case (ii) $\sqcap = \odot$ and $\Box = \odot$. The normality of μ_A immediately yields

$$(\mu_A \circ_\sqcap R_*)(y) = \bigvee_{x \in X} \{\mu_A(x) \odot [\mu_A(x) \odot \mu_B(y)]\} = \mu_B(y).$$

Case (iii) $\Box = \odot$ and $\sqcap = \wedge$. The normality of μ_A gives

$$(\mu_A \circ_\sqcap R_*)(y) = \bigvee_{x \in X} \{\mu_A(x) \wedge [\mu_A(x) \odot \mu_B(y)]\} = \mu_B(y).$$

Case (iv) $\Box = \wedge$ and $\sqcap = \odot$. Again using the normality of μ_A we obtain:

$$(\mu_A \circ_\sqcap R_*)(y) = \bigvee_{x \in X} \{\mu_A(x) \odot [\mu_A(x) \wedge \mu_B(y)]\} = \mu_B(y).$$

Case (v) $\sqcap = \wedge$ and $\Box = \wedge$. Finally, the normality of μ_A also ensures

$$(\mu_A \circ_\sqcap R_*)(y) = \bigvee_{x \in X} \{\mu_A(x) \wedge [\mu_A(x) \wedge \mu_B(y)]\} = \mu_B(y).$$

Exercise 2 For all counterexamples we assume $L = [0,1]$ endowed with Łukasiewicz structure.
(a) $\square = \rightarrow$ and $\sqcap = \wedge$. Let $X = \{x_1, x_2\}$, $Y = \{y_1, y_2\}$, $\mu_A(x_1) = 1$, $\mu_A(x_2) = 0.5$, $\mu_B(y_1) = 0$, $\mu_B(y_2) = 1$, $S(x_1, x_2) = 0.5$, $\mu_{A'}(x_1) = 1$, $\mu_{A'}(x_2) = 0$. Then we have $\widehat{(\mu_{A'} \circ_\sqcap R_*)}(y_1) = 0.5$, while $(\mu_{A'} \circ_\sqcap R_*)(y_1) = 0$.
(b) $\square = \odot$ and $\sqcap = \wedge$. Let $X = \{0, \frac{1}{2}, 1\}$, $Y = \{y_1, y_2\}$, $\mu_A(x) = x$, $\mu_B(y_1) = 1$, $\mu_B(y_2) = 0$, $S(x, x') = 1 - \min\{|x - x'|, 1\}$, $\mu_{A'}(0) = 1$, $\mu_{A'}(\frac{1}{2}) = \mu_{A'}(1) = 0$. Again we have $\widehat{(\mu_{A'} \circ_\sqcap R_*)}(y_1) = 0.5$, while $(\mu_{A'} \circ_\sqcap R_*)(y_1) = 0$.
(c) $\square = \sqcap = \wedge$. Let $X = \{x_1, x_2, x_3\}$, $Y = \{y_1, y_2\}$, $\mu_A(x_1) = 1$, $\mu_A(x_2) = 0.5$, $\mu_A(x_3) = 0$, $\mu_B(y_1) = 0$, $\mu_B(y_2) = 1$, $S(x_1, x) = \mu_A(x), x \in X$, $S(x_2, x_3) = 0.5$, $\mu_{A'}(x_1) = \mu_{A'}(x_2) = 0$, $\mu_{A'}(x_3) = 1$. Also in this case $\widehat{(\mu_{A'} \circ_\sqcap R_*)}(y_1) = 0.5$, while $(\mu_{A'} \circ_\sqcap R_*)(y_1) = 0$.

5.14 Exercises from Section 4.4

Exercise 1 Consider two fuzzy similarities (with respect to any BL-structure on the real unit interval!) S_1 and S_2 defined by

S_1	a	b	c
a	1	1	0
b	1	1	0
c	0	0	1

and

S_2	a	b	c
a	1	0	0
b	0	1	1
c	0	1	1

If one uses either Product structure or Gödel structure, then weak transitivity does not hold as in both cases

$$\begin{aligned} S\langle a,b\rangle \odot S\langle b,c\rangle &= \frac{1}{2}[(S_1\langle a,b\rangle + S_2\langle a,b\rangle)] \odot \frac{1}{2}[(S_1\langle b,c\rangle + S_2\langle b,c\rangle)] \\ &\not\leq \frac{1}{2}[(S_1\langle a,c\rangle + S_2\langle a,c\rangle)] \\ &= S\langle a,c\rangle \end{aligned}$$

Exercise 2 The described traffic situation has total similarity at the degree 1 with an `IF`-part
(wait = `verylong`) `AND` (vehicles = `some`) `AND` (gap = `large`).
The corresponding `THEN`-part is `red`.

Exercise 3 In this case the anaerobic threshold x beats/min has to fulfill $x \in [196 - 30, 196 - 10]$, therefore consider only the measured values 167, 172 , 180 and 187 beats/min. We have

Pulse	Blood Lac.inc.	Vent.inc.	Oxygen Up.inc.	Total sim.
166	0.8	1	0	0.6
172	1	0	1	0.67
180	1	0.5	0.34	0.61

The anaerobic threshold is 172 beats/min.

Exercise 4 Italy and France are more similar to each other that Finland and Belgium as Similar⟨Finland,Belgium⟩ = 0.92, Similar⟨Italy,France⟩ = 0.95.

Bibliography

1. L.P. Belluce, Generalized Fuzzy Connectives of MV-algebras. (to appear)

2. L.P. Belluce, Semisimple algebras of infinite valued logic and Bold fuzzy set theory, *Can. J. Math.* **38** (1986), 1356-1379.

3. L.P. Belluce, Semi-simple and complete MV-algebras, *Algebra Universalis* **29** (1992), 1-9.

4. G. Birkhoff, Latice Theory. 3rd ed. AMS Coll. Publ., Providence, N.Y. 1969.

5. J.L. Castro, F. Klawonn, Similarity in Fuzzy Reasoning, *Mathware and Soft Comp.* **2** (1995), 197-228.

6. C.C. Chang, Algebraic analysis of many-valued logics, *Trans. Amer. Math. Soc.* **88** (1958), 467-490.

7. C.C. Chang, A new proof of the completeness of Łukasiewicz axioms, *Trans. Amer. Math. Soc.* **93** (1959), 74-80.

8. R. Cignoli, I.M.L. D'Ottaviano, D. Mundici, Algebraic Foundations of Many-valued Reasoning. Kluwer Academic Publ., Dordrecht (to appear).

9. R.P. Dilworth, Abstract residuum over lattices, *Bull. Amer. Math. Soc.* **44** (1938), 262-268.

10. R.P. Dilworth, N. Ward, Residuated lattices, *Bull. Amer. Math. Soc.* **45** (1939), 335-354.

11. W.M. Faucett, Compact semigroups irreducible connected between two idempotents, *Proc. Amer. Math. Soc.* **6** (1955), 741-747.

12. J.M. Font, A.J. Rodriques, A. Torres, Wajsberg algebras, *Stochastica* **8**(1) (1984), 5-31.

13. B.R. Gaines, Foundations of fuzzy reasoning, *Int. Journal Man-Machine Stud.* **8** (1976), 623-668.

14. D. Gluschankof, Prime deductive systems and injective objects in the algebras of Łukasiewicz infinite-valued calculi, *Algebra Universalis* **29** (1992), 354-377.

15. P. Hájek, Metamathematics of fuzzy logic. Kluwer Academic Publ. Dordrecht. 1998.

16. P. Hájek, L. Godo and F. Esteva, A complete many-valued logic with product-conjunction. *Arch. Math. Logic* **35**(1996), 191-208.

17. M. Hempilä, Defining an athlete's aerobic and anaerobic thresholds with fuzzy total similarity. *Diploma Thesis.* Lappeenranta University of Technology, 1998.

18. C.S. Hoo, MV-algebras, ideals and semisimplicity, *Math. Japonica* **34** (1989), 563-583.

19. U. Höhle, Residuated l-monoids, in U. Hőhle and E.P. Klement (eds.) Non-classical Logics and Their Applications to Fuzzy Subsets: A Handbook of the Mathematical Foundations of Fuzzy Set Theory. Kluwer, Boston 1994.

20. L. Lacava, Alcune proprieta delle L-algebre e delle L-algebre esistenzialmente chiuse, *Bollettino U.M.I (5)* **16**-A (1979), 360-366.

21. L. Lacava, Sulle L-algebre iniettive, *Bollettino U.M.I (7)* **3**-A (1989), 319-324.

22. P. Mangani, Su certe algebre connesse con logiche a piu valori, *Bollettino U.M.I (4)* **8**-A (1973), 68-78.

23. J. Menu and J. Pavelka, A note on tensor products on the unit interval, *Comment. Math. Univ. Carol.* **17** (1976), 71-82.

24. P.S. Mostert and A.L. Shields, On the structure of semigroups on a compact manifold with boundary, *Annals of Math.* **65** (1957), 117-143.

25. J. Niittymäki and S. Kikuchi, Application of Fuzzy Logic to the Control of Pedestrian Crossing signal. *TRP Paper Number 980362.*

26. V. Novák, First-order fuzzy logic, *Studia Logica* **46**(1) (1989), 87-109.

27. J. Pavelka, On fuzzy logic, I,II,III *Zeitsch. f. Math. Logik* **25** (1979), 45-52, 119-134, 447-464.

28. H. Rasiowa, R. Sikorski, The Mathematics of Metamathematics, PWN, Warszawa, 1963.

29. P. Smets, P. Magres, The measure of the degree of truth and the grade of membership, *Int. Jour. Approximate Reasoning* **1** (1987), 327-347.

30. P. Smets, P. Magres, Implication in fuzzy logic, *Fuzzy Sets and Systems* **25** (1988), 67-72.

31. E. Trillas, L. Valverde, On implication and indistinguishability in the setting of fuzzy logic, in R.R. Jager and J. Kacprzyk (eds.) *Management Decision Support Systems using Fuzzy Set and Possibility Theory*, Verlag TUV Rheinland, Köln, 1984.

32. E. Trillas, L. Valverde, On Mode and Implication in Approximate Reasoning, in M.M. Gupta, A. Kandel, W. Beudler, J.B. Kiszka (eds.) *Approximate Reasoning in Expert Systems*, Elsevier Science Publ., Amsterdam, 1985.

33. E. Turunen, Algebraic structures in fuzzy logic, *Fuzzy Sets and Systems* **52** (1992), 181-188.

34. E. Turunen, On Generalized Fuzzy Relational Equations: Necessary and Sufficient Conditions for the Existence of Solutions, *Acta Univ. Carolin. Math. Phys.* **28** (1987), 33-37.

35. E. Turunen, Rules of inference in fuzzy sentential logic, *Fuzzy Sets and Systems.* **85**(1997) 63-72.

36. E. Turunen, Well-defined fuzzy sentential logic, *Math. Log. Quart.* **41** (1995), 236-248.

37. R.R. Yager, On a general class of fuzzy connectives, *Fuzzy Sets and Systems* **4** (1980), 235-242.

38. L.A. Zadeh, Fuzzy Sets, *Information and Control* **8** (1965), 338-353.

Index